Quick Arithmetic

A Self-Teaching Guide

Wiley Self-Teaching Guides teach practical skills from accounting to astronomy, management to mathematics. Look for them at your local bookstore.

Other Wiley Self-Teaching Guides:

Science

Astronomy, Fifth Edition, by Dinah L. Moche

Basic Physics, Second Edition, by Karl F. Kuhn

Biology, by Steven D. Garber

Chemistry: Concepts and Problems, Second Edition, by Clifford C. Houk and Richard Post

Geology, by Barbara Murck

Math

All the Math You'll Ever Need, Revised Edition, by Steve Slavin

Geometry and Trigonometry for Calculus, by Peter H. Selby

Practical Algebra, Second Edition, by Peter H. Selby and Steve Slavin

Quick Algebra Review, Second Edition, by Peter H. Selby and Steve Slavin

Quick Business Math, by Steve Slavin

Quick Calculus, Second Edition, by Daniel Kleppner and Norman Ramsey

Statistics, Fourth Edition, by Donald Koosis

Quick Arithmetic
A Self-Teaching Guide
Third Edition

Robert A. Carman

Santa Barbara Community College
Santa Barbara, California

Marilyn J. Carman

Santa Barbara City Schools
Santa Barbara, California

John Wiley & Sons, Inc.

New York • Chichester • Weinheim • Brisbane • Singapore • Toronto

This book is printed on acid-free paper. ⊗

Copyright © 1984 by John Wiley & Sons, Inc. All rights reserved
Copyright © 2001 by Robert A. Carman and Marilyn J. Carman

Published by John Wiley & Sons, Inc.
Published simultaneously in Canada

This publication is designed to provide accurate and authoritative information in regard to the subject matter covered. It is sold with the understanding that the publisher is not engaged in rendering professional services. If professional advice or other expert assistance is required, the services of a competent professional person should be sought.

Library of Congress Cataloging in Publication Data

Carman, Robert A.
    Quick arithmetic/Robert A. Carman, Marilyn J. Carman.—3rd ed.
        p. cm.
Includes index.
    ISBN 0-471-38494-1 (pbk.)
    1. Arithmetic. 2. Arithmetic—Programmed instruction. I. Carman, Marilyn J. II. Title.
    QA107 .C375 2001
    513'.07'7—dc21
                                                                      00-051336

10 9 8 7 6 5 4 3 2

For Ed Graper

Of making books
there is no end. . . .
This one's for you,
our guide, philosopher,
and friend.

Contents

Preface

Many very bright and competent people enrolled in colleges, universities, and community colleges are frustrated. They are eager, ambitious, and quite capable of succeeding in their careers or moving to a better job. They want to learn but find themselves handicapped because they do not have the basic mathematics skills needed to continue. They need help with these essential skills. If that describes where you are, this book is for you.

This book is designed to help you review or relearn basic arithmetic skills. It is more like a private tutor than a lecturer; you participate in the process rather than simply reading, listening, or sleeping through it.

The book is organized in a format that respects your unique needs and interests and teaches you accordingly:

- You can use it for **self-study**, for study with a tutor or helper, or as a text in a formal course.
- Each chapter begins with a **preview** and a sample test to help you see your particular needs.
- You have the **option** of designing your own course, skipping familiar material to save time or working through all of it if you need it.
- Many **practice problems** and self-tests are included, including drill problems, practical applications, more difficult brain boosters, and problems where a calculator should be used. Each chapter ends with an optional **self-test**.
- **Answers** to all problems are in the back of the book.

- Unlike previous mathematics textbooks you may have used, this book is careful to **explain** every operation. Sometimes we even explain our explanations.

This book has been used by hundreds of thousands of students and they tell us it is helpful, interesting, and even fun to work through. We hope you agree with them.

It is a pleasure for us to acknowledge our debts to the many people who have contributed to the development of this book and to this third edition. Jeffrey Golick and the staff at John Wiley & Sons, Inc., have been most supportive and patient throughout the lengthy process of producing a book. We were fortunate to have W. Royce Adams, formerly the director of the Reading Center at Santa Barbara Community College, read preliminary versions of the book and provide valuable assistance in improving its readability.

Finally, we wish to extend special thanks to our kindest critics and most enthusiastic helpers: our children—Pat, Laurie, Maire, and Eric— our other works in collaboration.

—RAC
—MJC

How to Use This Book

© 1970 United Feature Syndicate, Inc.

Many people go through life afraid of mathematics and upset by numbers. They bumble along miscounting their change, bouncing checks, and eventually trying to avoid college courses or jobs that require even simple math. Most such people need to return and make a fresh start. Few get the chance. This book presents fresh-start math. It is designed so that you can:

- Start at the beginning or wherever you need to start
- Work on only what you need to know
- Move as fast or as slowly as you wish
- Skip material you already understand
- Do as many practice problems as you need
- Take self-tests to measure your progress

In other words, if you find mathematics difficult and want a fresh start, this book is designed for you.

This is no ordinary book. You cannot easily browse in it. You don't read it; you work your way through it. Ideas are arranged step-by-step in short portions or frames. Each **frame** contains information, careful explanations, examples, and questions to test your understanding. Frames are numbered on the left.

7 Read the material in each frame carefully, follow the examples, and answer the questions that lead to the next frame. Correct answers move you quickly through the book. Incorrect answers lead you to frames providing further information. You move through the book frame by frame, sometimes forward, sometimes backward.

Each major section of the book starts with a **preview** that will help you determine those parts on which you need to work.

Notice the following symbols designed to help you:

CAUTION ▷ This symbol points out **common errors** or misunderstandings that many students have and about which you need to be careful.

LEARNING HELP ▷ This offers a **hint** or gives an **alternate explanation** or different way of thinking about a concept or procedure.

 This **calculator** icon tells you to look for a calculator key sequence showing how to solve the problem using a calculator. Calculators are an important tool, and we assume that once you have learned the basic operations of mathematics, you will use a calculator.

Important terms are noted in the margin where they are first used or defined.

Students are led step-by-step through examples and explanations:

- **Step 1** Many **worked examples** are given . . .
- **Step 2** . . . with **explanations** for each step . . .

• **Step 3** . . . and immediate feedback in the text and in Boxed Comments ➤

and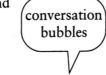

Most students hesitate to ask questions. They would rather risk failure than look foolish by asking "dumb" questions. To relieve this worry, we ask and answer these FAQs (frequently asked questions) in the cartoons. Learning the FAQs produces smart students.

Special Topics
As you move through the book, special topics appear within lines like this. Check them out.

In 1846, the Reverend H. W. Adams described what happened when the 10-year-old math whiz Truman Safford was asked to multiply, in his head, the number 365,365,365,365,365,365 by itself. "He flew around the room like a top, pulled his pantaloons over the tops of his boots, bit his hands, rolled his eyes in their sockets, sometimes smiling and talking, and then, seeming to be in agony, in not more than one minute, he said 133,491,850,208,566,925,016,658,299,941,583,225."* In this book we will show you a way to do arithmetic that is not so strenuous, quite a bit slower, and not nearly so much fun to watch.

Now, turn to page 1 and let's begin.

*James R. Newman, *The World of Mathematics* (New York: Simon and Schuster, 1956), p. 466.

1

Whole Numbers

PREVIEW 1

	Where to Go for Help	
When you successfully complete this chapter you will be able to do the following:	Page	**Frame**

1. Add, subtract, multiply, and divide whole numbers.

	Page	Frame
(a) $6341 + 14{,}207 + 635 =$ _____	3	**1**
(b) $64{,}508 - 37{,}629 =$ _____	21	**16**
(c) $4328 \times 417 =$ _____	32	**24**
(d) $672 \times 2009 =$ _____	32	**24**
(e) $46{,}986 \div 745 =$ _____	46	**35**
(f) $37\overline{)3003} =$ _____	46	**35**
(g) $\dfrac{1541}{23} =$ _____	46	**35**
(h) $12 \times 0 =$ _____	32	**24**
(i) $16 \div 1 =$ _____	46	**35**

Where to Go for Help

	Page	Frame

2. Write a whole number as a product of its prime factors.

(a) $3780 =$ _____   57   **44**

(b) $1848 =$ _____   57   **44**

3. Calculate integer powers of a whole number.

(a) $2^3 =$ _____   72   **59**

(b) $42^2 =$ _____   72   **59**

4. Find the square root of a perfect square.

(a) $\sqrt{169} =$ _____   72   **59**

If you are certain you can work all of these problems correctly, turn to page 81 for a self-test. If you want help with any of these objectives or if you cannot work one of the preview problems, turn to the page indicated. Super-students (those who want to be certain they learn all of this), turn to frame **1** and begin work there.

ANSWERS TO PREVIEW 1 PROBLEMS

1. (a) 21,163
(b) 26,879
(c) 1,804,776
(d) 1,350,048
(e) 63 with remainder 51
(f) 81 with remainder 6
(g) 67
(h) 0
(i) 16

2. (a) $2^2 \cdot 3^3 \cdot 5 \cdot 7$
(b) $2^3 \cdot 3 \cdot 7 \cdot 11$

3. (a) 8
(b) 1764

4. (a) 13

---

1 WHOLE NUMBERS

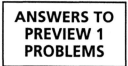

©1970 United Feature Syndicate, Inc.

1 Reading and Writing Numbers

1 Once upon a time, most people used numbers to tell time, count things, and keep track of their lunch money. But we are living in an age of calculators and computers, and math skills are important in everything we do. Using a calculator requires a good understanding of basic math skills and concepts. For most educated adults, working with numbers is as important a part of their job as being able to read and write.

In this chapter we will take a how-to-do-it look at the basic operations of mathematics: addition, subtraction, multiplication, and division.

What is a number? It is a way of thinking, an idea, that enables us to compare very different sets of objects. It is the idea behind the act of counting. The number three is the idea that describes any collection of three objects: 3 people, 3 trees, 3 colors, 3 dreams. We recognize that these collections all have the quality of "threeness" even though they may differ in every other way.

Numerals We use **numerals** to name numbers. For example, the number of corners on a square is four, or 4, or IV in Roman numerals, or 囚 in Chinese numerals.

Digits In our modern number system we use ten **digits**—0, 1, 2, 3, 4, 5, 6, 7, 8, and 9—to build numerals just as we use the twenty-six letters of the alphabet to build words.

Is 10 a digit? Think about it. Then turn to **3** to continue.

2 Hi. What are you doing here? Lost? Window shopping? Just passing through? Nowhere in this book are you directed to frame **2**. (Notice that little **2** to the left above? That's a frame number.) Remember, in this book you move from frame to frame as directed, but not necessarily in 1-2-3 order. Follow directions and you'll never get lost.

Now return to **1** and keep working.

3 No, 10 is not a digit. It is a numeral formed from the two digits 1 and 0. Remember:

- A **number** is an idea related to counting.
- A **numeral** is a symbol used to represent a number.
- A **digit** is one of the ten symbols (0, 1, 2, 3, 4, 5, 6, 7, 8, 9) we use to form numerals.

How many letters are in this set?

Count them. Write your answer, and then turn to frame **4**.

4   We counted 23, and of course we write it in the ordinary, everyday manner. Leave the Roman, Chinese, and other numeral systems to Romans, Chinese, and people who enjoy the history of mathematics.

The basis of our system of numeration is grouping into sets of ten or multiples of ten.

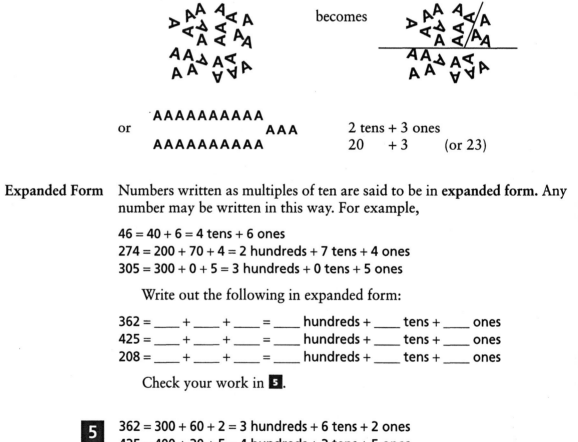

becomes

or   AAAAAAAAAA
                                    AAA   2 tens + 3 ones
            AAAAAAAAAA   20   + 3   (or 23)

Expanded Form   Numbers written as multiples of ten are said to be in **expanded form**. Any number may be written in this way. For example,

46 = 40 + 6 = 4 tens + 6 ones
274 = 200 + 70 + 4 = 2 hundreds + 7 tens + 4 ones
305 = 300 + 0 + 5 = 3 hundreds + 0 tens + 5 ones

Write out the following in expanded form:

362 = ____ + ____ + ____ = ____ hundreds + ____ tens + ____ ones
425 = ____ + ____ + ____ = ____ hundreds + ____ tens + ____ ones
208 = ____ + ____ + ____ = ____ hundreds + ____ tens + ____ ones

Check your work in **5**.

5   362 = 300 + 60 + 2 = 3 hundreds + 6 tens + 2 ones
425 = 400 + 20 + 5 = 4 hundreds + 2 tens + 5 ones
208 = 200 + 0 + 8 = 2 hundreds + 0 tens + 8 ones

Place Value Notice that the 2 in 362 means something very different from the 2 in 425 or 208. In 362 the 2 signifies two ones. In 425 the 2 signifies two tens. In 208 the 2 signifies two hundreds. Ours is a **place value** system of naming numbers: the value of any digit depends on the place where it is located.

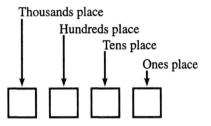

Writing numbers in expanded form will be helpful later when you want to understand how arithmetic operations work.

Naming Large Numbers

Any large number given in numerical form may be translated to words by using the following diagram.

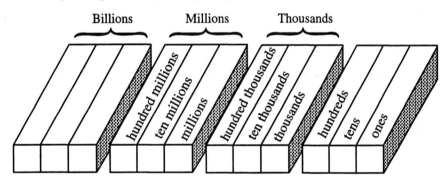

The number 14,237 can be placed in this diagram like this

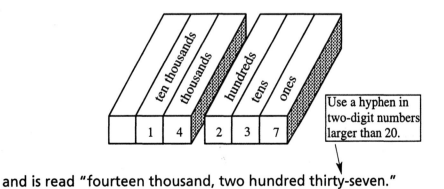

Use a hyphen in two-digit numbers larger than 20.

and is read "fourteen thousand, two hundred thirty-seven."

The number 47,653,290,866 becomes

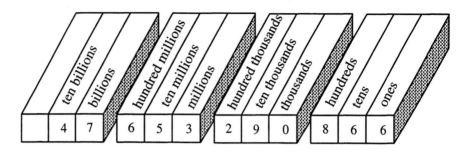

and is read "forty-seven billion, six hundred fifty-three million, two hundred ninety thousand, eight hundred sixty-six."

In each block of three digits read the digits in the normal way ("forty-seven," "six hundred fifty-three") and add the name of the block ("billion," "million"). Notice that the word "and" is not used in naming these numbers.

Name each of the following numbers in words.

1. 12,043
2. 457,009
3. 2,023,067
4. 102,400
5. 4,005,007
6. 342,103,010
7. 34,002
8. 82,004,700
9. 2,007,054,002

The correct answers are in the Appendix.

Roman Numerals

A number is an idea. A **numeral** is a symbol that enables us to express that idea in writing and use it in counting and calculating. Roman numerals were used by the ancient Romans almost 2000 years ago and are still seen on clock faces, building inscriptions, and textbooks. The following seven symbols are used:

I	V	X	L	C	D	M
1	5	10	50	100	500	1000

Notice that the numbers represented are 1, 5, and multiples of 5 and 10 (the number of fingers on one hand and on two hands). There is no zero. We write numerals with these symbols by placing them in a row and adding or subtracting. For example,

1 = I	7 = VII (V + I + I)
2 = II	8 = VIII
3 = III	9 = IX (I subtracted from X)
4 = IV (I subtracted from V)	10 = X
5 = V	27 = XXVII
6 = VI (V + I)	152 = CLII

The Romans used only addition and they wrote 4 as IIII, but in order to keep numerals smaller, later mathematicians used subtraction to form numbers like these:

IV	IX	XL	XC	CD	CM
4	9	40	90	400	900

Only these six subtractions are allowed. From these other combinations can be made.

XIX = X + IX or 10 + 9 or 19
LXIV = LX + IV or 60 + 4 or 64

Roman numerals are a bit more difficult to write than the ones we use and they are a headache to multiply or divide, but they are very easy to add or subtract. For example, 111 + 16 = 127 would be written like this:

CXI + XVI = CXXVII

The numerals we use now (0, 1, 2, 3, etc.) were first seen in Europe in about the thirteenth century, but Roman numerals were used by bankers and bookkeepers until the eighteenth century. They did not trust symbols like 0 that could easily be changed to 6, 8, or 9 by a dishonest clerk.

Translate the following numbers written in Roman numerals into modern numerals.

1. XIII	**2.** XVII	**3.** XXVIII
4. LXXVII	**5.** CXXIV	**6.** CDXXXI
7. MDCCCXCII	**8.** MCMX	**9.** MCMXXXIV
10. MCMLXVI		

The answers are in the Appendix.

Turn to frame **6** .

2 Adding Whole Numbers

6 Addition is the simplest arithmetic operation.

4   +   3   =   7

We add collections of objects by combining them into a single set and then counting and naming that new set. The numbers being added (4 and 3 in this case) are called **addends** and 7 is the **sum** of the addition.

There are a few simple addition facts you should have stored in your memory and ready to use. Complete the following table by adding the number at the top to the number at the side and placing their sum in the proper square. We have added 1 + 2 = 3 and 4 + 3 = 7 for you.

Add	4	2	8	7	5	6	1	3	9
2									
4								7	
7									
5									
1		3							
9									
6									
8									
3									

Check your answer in **7**.

7 Here is the completed addition table:

Add	4	2	8	7	5	6	1	3	9
2	6	4	10	9	7	8	3	5	11
4	8	6	12	11	9	10	5	7	13
7	11	9	15	14	12	13	8	10	16
5	9	7	13	12	10	11	6	8	14
1	5	3	9	8	6	7	2	4	10
9	13	11	17	16	14	15	10	12	18
6	10	8	14	13	11	12	7	9	15
8	12	10	16	15	13	14	9	11	17
3	7	5	11	10	8	9	4	6	12

Did you notice that changing the order in which you add numbers does not change their sum?

$4 + 3 = 7$ and $3 + 4 = 7$
$2 + 4 = 6$ and $4 + 2 = 6$

This is true for any addition problem involving whole numbers. It is known as the **commutative property of addition**. (A commuter is a person who changes location daily, moving back and forth between suburbs and city. The commutative property says changing the location or order of the numbers being added does not change their sum.)

If you have not already memorized the addition of one-digit numbers, it is time to do so. To help you, a study card for addition facts is provided in the back of this book in the Appendix. Use it if you need it.

If you want more practice adding one-digit numbers, go to **9**. Otherwise, continue in **8**.

8 Now, let's try a more difficult addition problem.

$34 + 52 = $ _____

Estimating   The first step in any arithmetic problem is to **estimate** or **guess** at the answer. Never work a problem until you know roughly what the answer is going to be. Always know where you are going.

Make a guess at the answer to the problem on page 9. Write your guess here _____ and then turn to 🔟 and continue.

9   *Problem Set 1-1:*   Practice Problems for One-Digit Addition
Add the following. Work quickly. You should be able to answer all problems in a set correctly in the time indicated. (The times are for community college students enrolled in a developmental math course.) Try to do all addition mentally. The answers are in the Appendix.

A. Add (average time: 90 seconds):

3	7	3	8	5	3	9	2	6	8
5	9	3	5	6	8	4	7	7	4

7	9	4	7	8	6	7	9	8	5
7	8	2	5	7	3	4	3	8	4

9	2	6	2	8	4	2	5	9	8
6	8	4	9	6	3	6	9	9	7

8	5	7	8	9	6	9	7	8	6
6	9	5	4	7	6	4	6	5	9

7	8	6	3	8	7	6	4	9	5
4	9	5	9	3	8	7	8	9	7

B. Add (average time: 90 seconds):

7	5	2	5	8	4	3	8	9	7
3	6	9	7	8	5	6	4	3	6

6	8	9	3	7	2	9	9	7	4
4	5	6	5	7	7	4	9	8	7

8	9	8	5	8	4	9	6	4	8
3	7	6	5	9	5	5	6	3	2

5	6	7	7	5	6	2	3	6	9
8	7	5	9	4	5	8	7	8	8

7	5	9	4	3	8	7	8	5	7
4	9	2	6	8	6	9	4	8	8

C. Add (average time: 90 seconds):

2	7	3	4	2	6	3	5	9	5
5	3	6	5	7	7	4	7	6	2
4	2	5	8	9	8	4	8	3	8

6	5	4	8	6	9	7	4	8	1
2	4	2	1	8	3	1	9	4	8
7	5	9	9	8	5	6	1	6	7

1	9	3	1	7	2	9	9	8	5
9	9	1	6	9	9	8	5	3	4
2	1	4	3	6	1	2	1	3	7

Turn to frame **8**.

10 34 + 52 is approximately 30 + 50 or 80. The correct answer will be about 80, not 8 or 800 or 8000. Once you have a reasonable **estimate** of the answer, you are ready to do the arithmetic work.

34 + 52 = ____

Check your answer in **11**.

11 You should have set the problem up like this:

> **Step 1** The numbers to be added are arranged vertically in columns.
> **Step 2** The right end or ones digits are placed in the ones column, the tens digits are placed in the tens column, and so on.

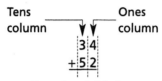

Tens   Ones
column   column

$$\begin{array}{r} 3\,4 \\ +\,5\,2 \end{array}$$

> **Step 3** Add the numbers in each column separately.

For the ones   For the tens
digits: digits:

$$\begin{array}{r} 3\,4 \\ +\,5\,2 \\ \hline 6 \end{array} \quad 4+2=6 \qquad\qquad \begin{array}{r} 3\,4 \\ +\,5\,2 \\ \hline 8\,6 \end{array} \quad 3+5=8$$

CAUTION ⇨ The most frequent cause of errors in arithmetic is carelessness, especially in simple procedures such as lining up the digits correctly.
Avoid the confusion of

$$\begin{array}{r} 34 \\ +\ 52 \\ \hline \end{array} \qquad \text{or} \qquad \begin{array}{r} 34 \\ +52 \\ \hline \end{array}$$

Once the digits are lined up the problem is easy.

Does your answer agree with your original guess? Yes. The guess, 80, is roughly equal to the actual sum, 86.

What we have just shown you is the **guess 'n check** method of doing mathematics calculations.

- **Step 1** Guess at the answer. This is not a wild guess but a careful estimate.
- **Step 2** Work the problem carefully.
- **Step 3** Check your answer against your guess. If they are very different, repeat both step 1 and step 2.

Most students hesitate to guess, afraid they might guess incorrectly. Relax. You are the only one who will know your guess. Do it in your head, do it quickly, and make it reasonably accurate. Step 3 helps you detect incorrect answers before you finish the problem. The guess 'n check method means you never work in the dark, you always know where you are going. Use this approach on every math calculation and you need never have an incorrect answer again.

Guess 'n check is especially important when you use a calculator.

Here is a slightly more difficult problem. Try it, then go to **12**.

$$27 + 48 = \underline{\hspace{1cm}}$$

---

12 First, **guess.** Estimate the answer.

27 + 48 is roughly 30 + 50 or about 80. The answer is closer to 80 than to 8 or 800.

Second, line up the addends vertically.

$$\begin{array}{r} 27 \\ + 48 \end{array}$$

Third, work it out carefully.

$$\begin{array}{r} \overset{1}{2}7 \\ + 48 \\ \hline 75 \end{array}$$

Finally, **check** your answer against your guess.

The guess, 80, is roughly equal to the actual answer, 75. We write this as 75 ≈ 80. The wiggly equals sign means "approximately equals."

What does that little 1 above the tens column mean? What really happens when you "carry" a digit? In expanded notation:

$$\begin{array}{l} 27 = 2 \text{ tens} + 7 \text{ ones} \\ + 48 = 4 \text{ tens} + 8 \text{ ones} \\ \hline 6 \text{ tens} + 15 \text{ ones} = 6 \text{ tens} + 1 \text{ ten} + 5 \text{ ones} \\ \phantom{+48 = 6 \text{ tens} + 15 \text{ ones} =} 7 \text{ tens} + 5 \text{ ones} \\ \phantom{+48 = 6 \text{ tens} + 15 \text{ ones} =} = 75 \end{array}$$

The 1 that is "carried" over to the tens column is really a 10! Another way to see this is shown on page 14.

Step 1	Step 2	Step 3

Step 1

    2 7
  + 4 8   add ones
    15   7 + 8 = 15

15 is the first
partial sum.

Step 2

    2 7
  + 4 8
    1 5   add tens
    6 0   20 + 40 = 60

60 is the second
partial sum.

Step 3

    27
  + 48
    15
    60   add partial sums
    75   15 + 60 = 75

75 is the final sum.

Again, you should see that the "carry 1" is the 10 in 15.

Shortcut   Using partial sums is the long way to add, so usually we take a shortcut and write:

     1
    27   7 + 8 = 15
  + 48   Write 5,
     5   carry 1 ten.

     1
    27
  + 48
    75   1 + 2 + 4 = 7

You will learn the shortcut method here, but it is important that you know why it works.

Use the partial sum method to work this problem.

429 + 758 = _____

Set the problem up step by step as we did above, then turn to **13**. (Don't forget to guess 'n check.)

I still count on my fingers when I add. Is that bad?

Not really. It's slow and sometimes inconvenient, but it is the normal way to start. Keep working to memorize one-digit number additions.

13 429 + 758 = _____

Guess: 400 + 800 = 1200

Line up the addends: 429
                                  + 758

Use partial sums: 429
                                  + 758
                                    17   add ones: 9 + 8 = 17
                                    70   add tens: 20 + 50 = 70
                                  1100   add hundreds: 400 + 700 = 1100
                                  1187   add the partial sums

Check: 1200 (the guess) roughly equals 1187.   1200 ≈ 1187.

And of course you would use the shortcut method once you understand this process.

Step 1   **Step 2**   **Step 3**
  1   1   1
  429   9 + 8 = 17   429   429
+ 758   Write 7,   + 758   1 + 2 + 5 = 8   + 758   4 + 7 = 11
    7   carry 1 ten   87   1187

Now try these problems. Work them first using partial sums, then using the shortcut. Be sure to guess 'n check.

(a) 246 + 877 = _____

(b) 2348 + 724 = _____

(c) 980 + 436 = _____

Check your answers in **14** when you are finished.

How to Add Long Lists of Numbers

Very often, especially in business and industry, it is necessary to add long lists of numbers. The best procedure is to break the problem down into a series of simpler additions. First add sets of two or three numbers, then add these sums to obtain the total. For example,

```
  9
  3 ----- 12
  7
  6     13      25
 12
  4     16
 17
 +5 ----- 22 ----- 38
                   63
```

Using this procedure you do a little more writing but carry fewer numbers in your head. The result is fewer mistakes.

Better yet, keep your eye open for combinations that add to 10 or 15, and work with mental addition of three addends.

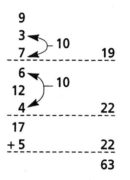

```
  9
  3
  7      10        19
  6
 12      10
  4                22
 17
 +5                22
                   63
```

Try these problems for practice.

1.	2.	3.	4.	5.	6.
8	7	3	11	3	13
17	6	5	7	5	17
3	8	7	2	12	11
4	5	6	5	7	14
11	9	5	6	6	15
9	3	1	7	4	8
16	7	3	13	1	9
7	12	4	6	2	16
11	8	2	5	18	12
5	16	7	14	9	7
		3	16	7	18

The answers are in the Appendix.

14   (a) Guess: $200 + 900 = 1100$

$$\begin{array}{r} 246 \\ + 877 \\ \hline \end{array}$$

13	add ones: $6 + 7 = 13$
110	add tens: $40 + 70 = 110$
1000	add hundreds: $200 + 800 = 1000$
1123	add partial sums

Check: 1100 is roughly equal to 1123. $1100 \approx 1123$.

Shortcut method:

Step 1

$$\begin{array}{r} \overset{1}{2}46 \\ + 877 \\ \hline 3 \end{array}$$ $6 + 7 = 13$
Write 3,
carry 1 ten.

Step 2

$$\begin{array}{r} \overset{1\;1}{2}46 \\ + 877 \\ \hline 23 \end{array}$$ $1 + 4 + 7 = 12$
Write 2,
carry 1 hundred.

Step 3

$$\begin{array}{r} \overset{1\;1}{2}46 \\ + 877 \\ \hline 1123 \end{array}$$ $1 + 2 + 8 = 11$

(b) Guess: $2300 + 700 = 3000$

$$\begin{array}{r} 2348 \\ + 724 \\ \hline \end{array}$$

12	add ones: $8 + 4 = 12$
60	add tens: $40 + 20 = 60$
1000	add hundreds: $300 + 700 = 1000$
2000	add thousands: 2000
3072	add partial sums

Check: The guess, 3000, is roughly equal to 3072. $3000 \approx 3072$.

Shortcut method:

Step 1

$$\begin{array}{r} 2\overset{1}{3}48 \\ + 724 \\ \hline 2 \end{array}$$ $8 + 4 = 12$
Write 2,
carry 1 ten.

Step 2

$$\begin{array}{r} 2\overset{1}{3}48 \\ + 724 \\ \hline 72 \end{array}$$ $1 + 4 + 2 = 7$
Write 7.

Step 3

$$\begin{array}{r} 2\overset{1}{3}48 \\ + 724 \\ \hline 072 \end{array}$$ $3 + 7 = 10$
Write 0,
carry 1 thousand.

Step 4

$$\begin{array}{r} \overset{1\;\;1}{2}348 \\ + 724 \\ \hline 3072 \end{array}$$ $1 + 2 = 3$

(c) Guess: 1000 + 400 = 1400

$$
\begin{array}{r}
980 \\
+\,436 \\
\end{array}
$$

6	add ones: 0 + 6 = 6
110	add tens: 80 + 30 = 110
1300	add hundreds: 900 + 400 = 1300
1416	

Check: 1400 is approximately equal to 1416.   1400 ≈ 1416.

Shortcut method:

Step 1

$$
\begin{array}{r}
980 \\
+\,436 \\
\hline
6
\end{array}
$$ 0 + 6 = 6

Step 2

$$
\begin{array}{r}
\overset{1}{9}80 \\
+\,436 \\
\hline
16
\end{array}
$$ 8 + 3 = 11
Write 1,
carry 1 hundred.

Step 3

$$
\begin{array}{r}
\overset{1}{9}80 \\
+\,436 \\
\hline
1416
\end{array}
$$ 1 + 9 + 4 = 14

If you had difficulty with any of these problems, you should return to **5** and review. Otherwise go to **15** for a set of practice addition problems.

15   *Problem Set 1-2:*   Adding Whole Numbers

A. Add:

1. 47	2. 18	3. 27	4. 57	5. 45	6. 89
23	86	38	69	35	17

7. 73	8. 44	9. 92	10. 38	11. 88	12. 75
39	28	39	65	17	48

13. 47	14. 26	15. 76	16. 48	17. 33	18. 67
56	98	24	84	19	69

B. Add:

1. 273	2. 189	3. 726	4. 508	5. 701
142	204	387	495	829

6. 684	7. 729	8. 432	9. 708	10. 621
706	287	399	554	388

11. 386	12. 747	13. 593	14. 375	15. 906
438	59	648	486	95

C. Add:

1. 4237	2. 6489	3. 5076	4. 1684
1288	3074	4385	927

5. 7907	6. 1467	7. 3015	8. 9864
1395	2046	687	2735

9. 6872	10. 8360	11. 6009	12. 3785
493	1762	496	7643

13. 5049	14. 6709	15. 8475	16. 6008
732	9006	928	5842

D. Add:

1. 18,745	2. 10,674	3. 60,485	4. 12,008
6,972	397	9,766	9,634

5. 9,876	6. 59,684	7. 40,026	8. 78,044
4,835	29,527	7,085	97,684

9. 94,036	10. 87,468
6,975	92,729

E. Arrange vertically and add:

1. 487 + 29 + 526 =

2. 715 + 4293 + 184 + 19 =

3. 1706 + 387 + 42 + 307 =

4. 456 + 978 + 1423 + 3584 =

5. 6284 + 28 + 674 + 97 =

6. 6842 + 9008 + 57 + 368 =

7. 322 + 46 + 5984 =

8. 7268 + 209 + 178 =

9. 5016 + 423 + 1075 =

10. 8764 + 85 + 983 + 19 =

F. Brain Boosters   Brain Boosters are more difficult and more fun than the regular problems. You will find them challenging, but don't expect to be as successful with them as you are with the others.

1.   The Aero University kite flying team weighs in as follows: "Tank" Murphy, 263 lb; Bertha Brown, 218 lb; "Moose" Green, 314 lb; and Head Kiter, "Tiger" Smith, 87 lb. What is the total weight of the Mighty Kiters?

2.   At lunch the other day, John the calorie counter ate the following: one slice of whole wheat bread, 55 calories; cream cheese and honey on the bread, 148 calories; yogurt, 123 calories; fresh blackberries, 45 calories. What was his total calorie count for the meal?

3.   **Business**   During the first three months of the year, Balloons.com reported the following sales:

January	$613,572
February	$782,716
March	$834,247

What was their sales total for this quarter of the year?

4.   Which sum is greater?

987654321	123456789
87654321	123456780
7654321	123456700
654321	123456000
54321	123450000
4321	123400000
321	123000000
21	120000000
1	100000000

5.   **Building Construction**   The Happy Helper building materials supplier has four piles of bricks containing 1250, 865, 742, and 257 bricks. What is the total number of bricks on hand?

6.   **Business**   Micro Systems Support had the following income in the second quarter of the year: $23,572 in April, $22,716 in May, and $24,247 in June. What was its total income for this quarter?

G. Calculator Problems   The following problems are designed to be solved using a calculator. Be careful to estimate the answer before doing the calculation.

1. 437,049	**2.** 27,760	**3.** 13,907
361,204	932	20,764
378,991	40,087	127,045
+ 200,587	+ 35,268	+ 390,088

4. 4176 + 32 + 1007 + 4254 + 23,144 + 9006 =

5. 840 + 9801 + 32,009 + 47,606 + 966 + 3444 =

6. 2007 + 207 + 20,007 + 207 =

7. 5555 + 555 + 55 + 5555 + 55,555 =

8. 607 + 6607 + 6077 + 67 + 66,077 + 600,777 =

9. Office Services   Joe's Air Conditioning Company has not been very successful, and Joe is wondering if he should sell it and move to a better location. During the first six months of the year his expenses were:
Rent $1860	Taxes $315	Supplies $2540
Advertising $250	Part-time helper $2100	Transportation $948
Miscellaneous $187		

His income was:
January $609	February $1151	March $1269
April $1381	May $1687	June $1638

(a) What was his total expense for the six-month period?
(b) What was his total income for the six-month period?
(c) Rotate your calculator 180° to learn what Joe should do about this unhappy situation.

The answers to these problems are in the Appendix. When you have completed these practice problems you may continue in **16** with the study of subtraction or return to the preview for this chapter on page 1 and use it to determine the help you need next.

3   Subtracting Whole Numbers

16   Subtraction is the reverse of the process of addition.

Addition: 3 + 4 = □
Subtraction: 3 + □ = 7

Written this way, a subtraction problem asks the question, "How much must be added to a given number to produce a required amount?"

Most often, however, the numbers in a subtraction problem are written using a minus sign (−):

17 − 8 = □

This means that there is a number □ such that 8 + □ = 17. Write in the answer to this subtraction problem, then continue in **17**.

17

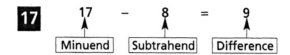

Special names are given to the numbers in a subtraction problem, and it will be helpful if you know them.

Minuend   The **minuend** is the larger of the two numbers in the problem. It is the number that is being decreased.

Subtrahend   The **subtrahend** is the number that is being subtracted from the minuend.

Difference   The **difference** is the amount that must be added to the subtrahend to produce the minuend. It is the answer to the subtraction problem.

The ability to solve simple subtraction problems depends on your knowledge of the addition of one-digit numbers. For example, find this difference:

9 − 4 = ____

Do this problem and then continue in **18**.

18   **9 − 4 = 5**

Solving this problem probably involved a chain of thoughts something like this:

"Nine minus four. Four added to what number gives nine? Five? Try it: four plus five equals nine. Right."

If you have memorized the addition of one-digit numbers (as shown in frame **6** or on the study card in the Appendix), subtraction problems involving small whole numbers will be easy for you. If you haven't memorized these, do it now.

Now try a more difficult subtraction problem.

54 − 23 = _____

What is the first step? Work the problem and continue in **19**.

19 The **first** step is to guess at the answer! Remember?

54 – 23 is roughly 50 – 20 or 30

The difference, your answer, will be about 30—not 3 or 300.

The **second** step is to write the numbers in a vertical format as you did with addition. Be careful to keep the ones digits in line in one column, the tens digits in a second column, and so on. Notice that the minuend is written above the subtrahend—larger number on top.

$$
\begin{array}{r}
5\,4 \\
-\,2\,3 \\
\hline
\end{array}
$$

Once the numbers have been arranged this way, the difference may be written immediately.

Step 1

$$
\begin{array}{r}
54 \\
-\,23 \\
\hline
1
\end{array}
$$
ones digits: 4 – 3 = 1

Step 2

$$
\begin{array}{r}
54 \\
-\,23 \\
\hline
31
\end{array}
$$
tens digits: 50 – 20 = 30

The difference is 31 and this agrees with our first guess: $30 \approx 31$.

Suppose we need to subtract 8 from 24.

24 – 8 = ___?___

Because 8 is larger than 4 we need to do some rearranging:

24 = 10 + 14

and 24 – 8 = 10 + 14 – 8 = 10 + (14 – 8)
= 10 + 6
= 16

Try this one:

64 – 37 = _____

Check your work in **20**.

20   **First,** guess. 64 – 37 is roughly 60 – 40 or 20. We estimate the answer to be about 20.

Second, arrange the numbers vertically in columns.

$$\begin{array}{r} 64 \\ -\ 37 \end{array}$$

Third, write them in expanded form to understand the process.

$$
\begin{array}{rclcl}
64 & = & 6 \text{ tens} + 4 \text{ ones} & = & 5 \text{ tens} + 14 \text{ ones} \\
-\ 37 & = & -(3 \text{ tens} + 7 \text{ ones}) & = & -(3 \text{ tens} + 7 \text{ ones}) \\
& & & = & 2 \text{ tens} + 7 \text{ ones} \\
& & & = & 20 \quad\quad + 7 \\
& & & = & 27 \text{ (which agrees with our guess)}
\end{array}
$$

Because 7 is larger than 4, we must "borrow" one ten from the six tens in the minuend. We are actually regrouping or rewriting the minuend.

In actual practice we do not write out subtraction problems in expanded form. Our work might look like this:

Step 1	Step 2	Step 3
		Borrow one ten,
64	$\overset{5}{\cancel{6}}\overset{14}{\cancel{4}}$ change the 6 in the	$\overset{5}{\cancel{6}}\overset{14}{\cancel{4}}$ 50 – 30 = 20,
– 37	– 37 tens place to 5,	– 37 write 2.
	7 change 4 to 14.	27
	subtract 14 – 7 = 7.	

LEARNING HELP ▷ Double-check subtraction problems by adding the answer and the subtrahend. Their sum should equal the minuend.

$$\begin{array}{r} 37 \\ + 27 \\ \hline 64 \end{array}$$

Try these problems for practice.

(a)
$$\begin{array}{r} 71 \\ - 39 \end{array}$$

(b)
$$\begin{array}{r} 283 \\ - 127 \end{array}$$

(c)
$$\begin{array}{r} 426 \\ - 128 \end{array}$$

(d)
$$\begin{array}{r} 902 \\ - 465 \end{array}$$

Solutions are in **21**.

Measurement Numbers

Many numbers are the result of measurement. A measurement number has two parts: a number part giving the size of the quantity and a unit giving a comparison standard for the measurement. For example, the winning time for a 100-yard dash is measured to be 9.62 seconds.

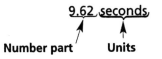

Number part   Units

Money numbers always have units: $5 or 25¢ are quantities measured in dollars and cents units.

To add or subtract measurement numbers:	Example:
1. Convert all numbers to the same units.	2 feet + 10 inches = 24 inches + 10 inches =
2. Add or subtract the number parts.	24 + 10 = 34
3. Attach the common units to the sum or difference.	34 inches

To multiply or divide measurement numbers:
1. Multiply or divide the number parts.
2. Attach the product or quotient of the units.

Example:

10 ft × 6 ft =

10 × 6 = 60

ft × ft = ft^2

10 ft × 6 ft = 60 ft^2

	Common English Units	Common Metric Units
Length or distance	inches, feet, miles	centimeters, meters, kilometers
Time	seconds, minutes, hours, days	seconds, minutes, hours, days
Weight or mass	ounces, pounds	grams, kilograms
Volume	cup, quart, gallon	cubic centimeter, liter
Area	square inch, square foot, acre	centimeter squared (cm^2), meter squared (m^2)

Metric Equivalents

1 meter	a little longer than a yard, about 1.09 yd
1 centimeter	the width of a paper clip, about 0.4 inches
1 kilometer	about 0.6 miles
1 liter	a little larger than one quart, about 1.06 qt
1 cubic centimeter	about one-fifth of a teaspoonful
1 kilogram	a little more than 2 pounds, about 2.2 lb

Practice working with measurement numbers by combining these quantities.

1. 23 ft + 7 ft = _____ ft

2. 46 lb − 18 lb = _____ lb

3. 3 ft 7 in. + 4 ft 8 in. = _____ ft _____ in.

4. 5 ft 8 in. − 2 ft 2 in. = _____ ft _____ in.

5. 8 lb 4 oz + 3 lb 13 oz = _____ lb _____ oz
   (Use 1 lb = 16 oz.)

6. 6 lb 12 oz − 3 lb 10 oz = _____ lb _____ oz

7. 4 hr 40 min − 2 hr 35 min = _____ hr _____ min

8. 16 cu ft + 12 cu ft = _____ cu ft

9. 43 sq in. − 17 sq in. = _____ sq in.

10. 64 mph − 36 mph = _____ mph

Check your answers in the Appendix.

---

21 (a) **Guess:** $70 - 40 = 30$

Step 1	**Step 2**	
	$\overset{6\,11}{7\cancel{1}}$	Borrow one ten from 70, change the 7 in the tens place to 6, change the 1 in the ones place to 11.
71		
− 39	− 39	
	32	$\boxed{11 - 9 = 2, \text{ write } 2}$
		$\boxed{60 - 30 = 30, \text{ write } 3}$

Check: The answer 32 is approximately equal to the guess, 30. $30 \approx 32$.

(b) **Guess:** $300 - 100 = 200$

Step 1	**Step 2**	
	$2\overset{7\,13}{8\cancel{3}}$	Borrow one ten from 30, change the 8 in the tens place to 7, change the 3 in the ones place to 13.
283		
− 127	− 127	
	156	$\boxed{13 - 7 = 6, \text{ write } 6}$
		$\boxed{70 - 20 = 50, \text{ write } 5}$
		$\boxed{200 - 100 = 100, \text{ write } 1}$

Check: The answer is approximately equal to the guess: $200 \approx 156$.

(c) **Guess:** $400 - 100 = 300$

Step 1	**Step 2**	**Step 3**	
	$4\overset{1\,16}{2\cancel{6}}$	$\overset{3\,11\,16}{\cancel{4}\cancel{2}\cancel{6}}$	$\boxed{16 - 8 = 8}$
426			$\boxed{110 - 20 = 90, \text{ write } 9}$
− 128	− 128	− 128	$\boxed{300 - 100 = 200, \text{ write } 2}$
	8	298	

Notice that in this case we must borrow twice. Borrow one ten from the 20 in 426 to make 16, then borrow one hundred from the 400 in 426 to make 110.

Check: $300 \approx 298$.

(d) Guess: $900 - 500 = 400$

Step 1	Step 2	Step 3

$$\begin{array}{r} 902 \\ -465 \\ \hline \end{array}$$

$$\begin{array}{r} {}^{8}{\cancel{9}}{}^{10}{\cancel{0}}2 \\ -465 \\ \hline \end{array}$$

$$\begin{array}{r} {}^{8}{\cancel{9}}{}^{9}{\cancel{0}}{}^{12}{\cancel{2}} \\ -465 \\ \hline 437 \end{array}$$

$\boxed{12 - 5 = 7}$
$\boxed{90 - 60 = 30, \text{ write } 3}$
$\boxed{800 - 400 = 400, \text{ write } 4}$

In this problem we first borrow one hundred from 900 to get a 10 in the tens place, then we borrow one ten from the tens place to get a 12 in the ones place. In expanded form problem (d) looks like this:

$$\begin{array}{r} 902 \\ -465 \\ \hline \end{array}$$

$$\begin{array}{r} 9 \text{ hundreds} + 0 \text{ tens} + 2 \text{ ones} \\ -(4 \text{ hundreds} + 6 \text{ tens} + 5 \text{ ones}) \\ \hline \end{array}$$

$$\begin{array}{r} 8 \text{ hundreds} + 10 \text{ tens} + 2 \text{ ones} \\ -(4 \text{ hundreds} + \ \ 6 \text{ tens} + 5 \text{ ones}) \\ \hline \end{array}$$

$$\begin{array}{r} 8 \text{ hundreds} + 9 \text{ tens} + 12 \text{ ones} \\ -(4 \text{ hundreds} + 6 \text{ tens} + \ \ 5 \text{ ones}) \\ \hline 4 \text{ hundreds} + 3 \text{ tens} + \ \ 7 \text{ ones} \end{array}$$

$$= 400 + 30 + 7$$

$$= 437$$

Check: $400 \approx 437$.

Do you want more worked examples of subtraction problems containing zeros, similar to this last one? If so, go to **23**. Otherwise go to **22** for a set of practice problems.

22 *Problem Set 1-3:* Subtracting Whole Numbers

A. Subtract:

1. 13	2. 9	3. 12	4. 15	5. 8	6. 13
7	4	5	9	6	8

7. 8	8. 7	9. 11	10. 6	11. 16	12. 16
0	7	7	5	7	8

13. 10	14. 13	15. 5	16. 18	17. 12	18. 10
7	6	5	9	9	3

19. 11
    8
	20. 5
	    1
		21. 10
		    2
			22. 8
			    7
				23. 14
				    6
					24. 13
					    9

25. 12
    3
	26. 9
	    2
		27. 15
		    6
			28. 11
			    4
				29. 9
				    0
					30. 12
					    8

31. 14
    5
	32. 9
	    3
		33. 9
		    6
			34. 1
			    0
				35. 11
				    5
					36. 14
					    8

37. 15
    7
	38. 11
	    6
		39. 12
		    7
			40. 17
			    9
				41. 13
				    6
					42. 15
					    8

43. 18
    0
	44. 14
	    7
		45. 16
		    9
			46. 7
			    4
				47. 12
				    4
					48. 17
					    8

B. Subtract:

1. 40
   27
	2. 63
	   19
		3. 78
		   49
			4. 33
			   17
				5. 51
				   39
					6. 85
					   28

7. 36
   17
	8. 60
	   43
		9. 42
		   27
			10. 91
			   63
				11. 52
				   16
					12. 47
					   29

13. 70
   48
	14. 94
	   57
		15. 34
		    9
			16. 55
			   29
				17. 56
				   18
					18. 93
					    8

C. Subtract:

1. 546
   357
	2. 640
	   182
		3. 409
		   324
			4. 914
			    37
				5. 476
				   195
					6. 219
					    43

7. 747
   593
	8. 564
	   298
		9. 400
		   127
			10. 316
			   118
				11. 803
				    88
					12. 327
					   276

13. 632
    58
	14. 525
	   480
		15. 438
		   409
			16. 701
			   556		

D. Subtract:

1.	6218	2.	8704	3.	6084	4.	30,209	5.	13,042
	3409		923		386		1,367		524

6.	8000	7.	57,022	8.	46,804	9.	5007	10.	10,785
	321		980		9,476		266		888

11.	10,000	12.	31,072	13.	48,093	14.	384,000	15.	27,004
	386		4,265		500		67,360		4,582

16.	60,754	17.	42,003
	5,295		17,064

E. **Brain Boosters**

1. The enrollment at Sunshine Tech is 5804. If 3985 students are men, how many female students are there?

2. Eric had a bad day at marbles. He started the day with 461 and arrived home after school with 177. How many marbles did he lose?

3. Last year on this date Sam's car odometer read 67,243 miles. It now reads 81,062 miles. How many miles has he driven in the past year?

4. If your income is $28,245 per year and you pay $4,658 in taxes, what is your take-home pay?

5. A $650 color TV is on sale for $495. How much money does Denny Doright save if he buys it at the sale price?

6. Subtract nine thousand, nine hundred nine from twelve thousand, twelve.

7. If you take three apples from a dish containing 13 apples, how many apples do you have?

8. On March 1, Mrs. Pennywatcher had a balance of $635 in her checking account. During March she deposited checks of $352 and $114 and wrote checks for $37, $216, $147, and $106. How much did she have left in her account at the end of the month?

9. Place "+" or "−" signs in each of the following sequences so that each one will total 100.

(a) 98 ____ 76 ____ 54 ____ 3 ____ 21 = 100

(b) 123 ____ 45 ____ 67 ____ 8 ____ 9 = 100

(c) 12 ____ 3 ____ 4 ____ 5 ____ 6 ____ 7 ____ 89 = 100

(d) 123 ____ 4 ____ 5 ____ 67 ____ 89 = 100

Can you make up more of these?

10. Is it true that 1963 pennies are worth almost $20?

F. Calculator Problems.   Solve using a calculator.

1. 46,804	2. 30,995	3. 669,406
– 999	– 6,887	– 80,507

4. 26,784 – 23,085 =

5. 237,895 – 188,096 =

6. 780,068 – 232,909 =

7. 1,210,110 – 897,068 =

8. Subtract 23,408 from 102,916.

9. Subtract 748,009 from 2,302,117.

10. The population of San Pablo county was 2,374,684 last year and increased to 2,405,505 this year. By what amount did the population grow?

The answers to the problems in Problem Set 1–3 are in the Appendix. When you have had the practice you need, go to **24** to study multiplication of whole numbers or return to the preview for this chapter on page 1.

23 Let's work through a few examples.

(a) **Step 1**	**Step 2**	**Step 3**	**Check:**
	³¹⁰	³⁹¹⁰	
400	4̸0̸0	4̸0̸0̸	167
– 167	– 167	– 167	+ 233
		233	400

Do you see in Step 3 that we have rewritten 400 as 300 + 90 + 10?

(b) Step 1

5006
− 2487

Step 2

⁴¹⁰
5̸0̸06
− 2487

Step 3

⁴⁹¹⁰
5̸0̸0̸6
− 2487

Step 4

⁴⁹⁹¹⁶
5̸0̸0̸6̸
− 2487
2519

Check:

2487
+ 2519
5006

(c) Step 1

24,632
− 5,718

Step 2

²¹²
24,63̸2̸
− 5,718
14

Step 3

³ ¹⁶²¹²
24,6̸3̸2̸
− 5,718
914

Step 4

¹¹³ ¹⁶²¹²
2̸4̸,6̸3̸2̸
− 5,718
18,914

Check:

5,718
+ 18,914
24,632

Any subtraction problem that involves borrowing should *always* be checked. It is very easy to make a mistake in the borrowing process.

Go to **22** for a set of practice problems on subtraction.

4   Multiplying Whole Numbers

24 In a certain football game, the West Newton Waterbugs scored five touchdowns at six points each. How many total points did they score through touchdowns? We can answer the question several ways.

1.   Count points: ·

2.   Add touchdowns: $6 + 6 + 6 + 6 + 6 = ?$

3.   Multiply: $5 \times 6 = ?$

We're not sure about the mathematical ability of the West Newton scorekeeper, but most people would multiply. Multiplication is a shortcut method of counting or doing repeated addition.

How many points did the Waterbugs score? Work it out one way or another, and then go to **25**.

25

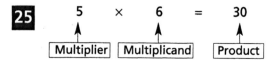

Multiplier
Product
Factors

In a multiplication statement the **multiplicand** is the number to be multiplied, the **multiplier** is the number multiplying the multiplicand, and the **product** is the result of the multiplication. The multiplier and multiplicand are called the **factors** of the product.

Notice that we can arrange these letters

a a a a
a a a a
a a a a
a a a a

into 3 rows of 4

a a a a
a a a a $3 \times 4 = 12$
a a a a

or 4 rows of 3

a a a
a a a $4 \times 3 = 12$
a a a
a a a

Changing the order of the factors does not change their product. This is the **commutative property of multiplication.**

In order to become skillful at multiplication, you must know the one-digit multiplication table from memory.

Complete the following table by multiplying the number at the top by the number at the side and placing their product in the proper square. We have multiplied $3 \times 4 = 12$ and $2 \times 5 = 10$ for you.

Multiply	2	5	8	1	3	6	9	7	4
1									
7									
5	10								
4				12					
9									
2									
6									
3									
8									

Check your answers in **27**.

The Sexy Six

Here are the six most often missed one-digit multiplications:

"Inside"
↓ ↓

$9 \times 8 = 72$ ⎫
$9 \times 7 = 63$ ⎬ →
$9 \times 6 = 54$ ⎭
$8 \times 7 = 56$
$8 \times 6 = 48$
$7 \times 6 = 42$

It may help you to notice that in these multiplications the "inside" digits (such as 8 and 7 in $9 \times 8 = 72$) are consecutive and the digits of the answer add to nine: $7 + 2 = 9$. This is true for *all* one-digit numbers multiplied by 9.

Be certain you have these memorized. (There is nothing very sexy about them, but we did get your attention, didn't we?)

---

26 Practice Problems: One-Digit Multiplication
Multiply as shown. Work quickly. You should be able to answer all problems in a set correctly in the time indicated. (These times are for community college students enrolled in a developmental math course.)

A. Multiply (average time: 100 seconds):

6	4	9	6	3	9	7	8	2	8
2	8	7	6	4	2	0	3	7	1

6	8	5	5	2	3	9	7	3	0
8	2	9	6	5	3	8	5	6	4

7	5	4	7	4	8	6	9	8	6
4	3	9	7	2	5	7	6	8	4

5	3	5	9	9	6	1	8	4	7
4	0	5	3	9	1	1	6	4	9

B. Multiply (average time: 100 seconds):

2	6	3	5	6	4	4	8	2	7
8	5	3	7	3	5	7	6	6	9

8	0	2	3	1	5	6	9	5	8
4	6	9	8	9	5	4	5	2	9

3	7	5	6	9	2	7	8	9	2
5	7	8	9	4	4	6	8	0	2

5	9	1	8	6	4	9	0	2	7
5	3	7	7	6	3	9	4	1	8

The answers to these problems are in the Appendix. When you have had the practice you need, turn to **28** and continue.

27 The completed multiplication table is shown below. If you are not able to perform these one-digit multiplications quickly from memory, you should practice until you can do so. A multiplication table is provided in the back of this book (page 268). Use it if you need it.

Multiply	2	5	8	1	3	6	9	7	4
1	2	5	8	1	3	6	9	7	4
7	14	35	56	7	21	42	63	49	28
5	10	25	40	5	15	30	45	35	20
4	8	20	32	4	12	24	36	28	16
9	18	45	72	9	27	54	81	63	36
2	4	10	16	2	6	12	18	14	8
6	12	30	48	6	18	36	54	42	24
3	6	15	24	3	9	18	27	21	12
8	16	40	64	8	24	48	72	56	32

Notice that the product of any number and 1 is that same number. For example,

$1 \times 2 = 2$
$1 \times 6 = 6$
$1 \times 753 = 753$

Zero has been omitted from the multiplication table because the product of any number and zero is zero. For example,

$0 \times 2 = 0$
$0 \times 7 = 0$
$0 \times 395 = 0$

If you want more practice in one-digit multiplication, go to **26**. Otherwise, go to **28**.

28 The multiplication of larger numbers is based on the one-digit number multiplication table. Find this product.

$34 \times 2 =$ _____

Remember the procedure you followed for addition. What are the first few steps in this multiplication? Try it, then go to .

29 **First, guess.** $30 \times 2 = 60$. The actual product of the multiplication will be about 60.

Second, arrange the factors to be multiplied vertically, with the ones digits in a single column, the tens digits in a second column, and so on.

Finally, to make the process clear, let's write it in expanded form.

$$\begin{array}{r} 34 \\ \times\, 2 \end{array} \longrightarrow \begin{array}{r} 3 \text{ tens} + 4 \text{ ones} \\ \times\, 2 \\ \hline 6 \text{ tens} + 8 \text{ ones} \end{array} = 60 + 8 = 68$$

Check: The guess 60 is roughly equal to the answer, 68.

Notice that when a single number multiplies a sum, it forms a product with each addend in the sum. For example,

$2 \times (30 + 4) = (2 \times 30) + (2 \times 4) = 60 + 8$

In the expanded multiplication above, the multiplier 2 forms a product with each addend in the sum (3 tens + 4 ones).

Write the following multiplication in expanded form.

$$\begin{array}{r} 28 \\ \times\, 3 \end{array}$$

Check your work in **30** .

Why do we use = for "equals"?

The mathematician Robert Recorde invented the equals sign in 1557. He decided that two parallel lines were as equal as anything available.

30 | **Guess:** $30 \times 3 = 90$

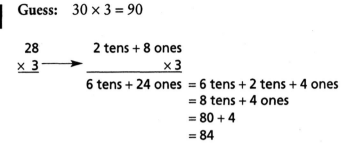

$$28 \qquad 2 \text{ tens} + 8 \text{ ones}$$
$$\times\, 3 \longrightarrow \qquad\qquad \times\, 3$$
$$\overline{6 \text{ tens} + 24 \text{ ones}} = 6 \text{ tens} + 2 \text{ tens} + 4 \text{ ones}$$
$$= 8 \text{ tens} + 4 \text{ ones}$$
$$= 80 + 4$$
$$= 84$$

Check: 90 is roughly equal to 84.

Of course we do not normally use the expanded form. Instead we simplify the work like this:

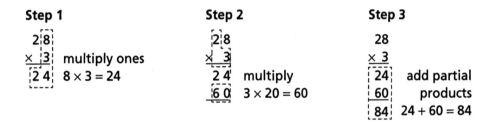

Step 1	**Step 2**	**Step 3**

Step 1

$2\,8$
$\times\ 3$ multiply ones
$\overline{2\,4}$ $8 \times 3 = 24$

Step 2

$2\,8$
$\times\ \ 3$
$\overline{2\,4}$ multiply
$6\,0$ $3 \times 20 = 60$

Step 3

28
$\times\ 3$
24 add partial
60 products
$\overline{84}$ $24 + 60 = 84$

When you are certain about how to do this, you can take a shortcut and write:

$\overset{2}{2}8$ $3 \times 8 = 24$, write 4 and carry 20.
$\times\ 3$ 3×2 tens = 6 tens, 6 tens + 2 tens = 8 tens, write 8.
$\overline{84}$

Try these problems to be certain you understand the process. Work all four problems using both the step-by-step and the shortcut methods.

(a)   43
       $\times\ 5$

(b)   73
       $\times\ 4$

(c)   29
       $\times\ 6$

(d)   287
       $\times\ \ 7$

Check your work in **31**.

31 (a) Guess: $5 \times 40 = 200$

Step 1	Step 2	Step 3

Step 1
```
  43
× 5     5 × 3 = 15
  15
```

Step 2
```
  43
×  5    5 × 40 = 200
  15
 200
```

Step 3
```
  43
×  5
  15
 200
 215
```

Shortcut

```
  43    5 × 3 = 15
×  5    5 × 4 = 20 tens
 215    20 tens + 1 ten = 21 tens
```

Check: $200 \approx 215$

(b) Guess: $70 \times 4 = 280$

Step 1
```
  73
× 4     3 × 4 = 12
  12
```

Step 2
```
  73
×  4    4 × 70 = 280
  12
 280
```

Step 3
```
  73
×  4
  12
 280
 292
```

Shortcut

$\overset{1}{7}3$ $4 \times 3 = 12$
$\times \quad 4$ $4 \times 7 = 28$ tens
$\overline{292}$ 28 tens + 1 ten = 29 tens

Check: $280 \approx 292$

(c) **Guess:** $30 \times 6 = 180$

Step 1	Step 2	Step 3
29	29	29
$\times\ 6$	$\times\ 6$	$\times\ 6$
54	54	54
	120	120
		174

Shortcut

$\overset{5}{2}9$ $6 \times 9 = 54$
$\times \quad 6$ 6×2 tens = 12 tens
$\overline{174}$ 12 tens + 5 tens = 17 tens

Check: $180 \approx 174$

(d) **Guess:** $300 \times 7 = 2100$

Step 1	Step 2	Step 3
287	287	287
$\times\ 7$	$\times\ 7$	$\times\ 7$
49 ◄ $\boxed{7 \times 7 = 49}$	49	49
	560 ◄ $\boxed{7 \times 80 = 560}$	560
		1400 ◄ $\boxed{7 \times 200 = 1400}$
		2009

Step 4

287
$\times \quad 7$
49
560
$\underline{1400}$
2009   Add

Shortcut

$\overset{6\,4}{287}$
$\times \quad 7$
$\overline{2009}$

Check: $2100 \approx 2009$

Calculations involving two-digit multipliers are done in exactly the same way. Apply this method to this problem.

$$\begin{array}{r} 57 \\ \times\, 24 \end{array}$$

The worked example is in **32**.

32   **Guess:** $60 \times 20 = 1200$

Step 1   **Step 2**

$$\begin{array}{r} 57 \\ \times\ 24 \\ \hline 28 \\ 200 \end{array}$$ Multiply by the ones digit (4).

$28 \leftarrow \boxed{4 \times 7 = 28}$

$200 \leftarrow \boxed{4 \times 50 = 200}$

$$\begin{array}{r} 57 \\ \times\ 24 \\ \hline 28 \\ 200 \\ 140 \\ 1000 \\ \hline 1368 \end{array}$$ Multiply by the tens digit (2).

$140 \leftarrow \boxed{20 \times 7 = 140}$

$1000 \leftarrow \boxed{20 \times 50 = 1000}$

Add

Check: $1200 \approx 1368$

Use the shortcut method illustrated next to reduce the written work.

$$\begin{array}{r} \overset{\overset{1}{2}}{57} \\ \times\ 24 \\ \hline 228 \\ 1140 \\ \hline 1368 \end{array}$$

$\boxed{\begin{array}{l} 4 \times 7 = 28, \text{ write 8, carry 2 tens.} \\ 4 \times 50 = 200, \text{ add carried 20 to get 220, write 22.} \end{array}}$

$\boxed{\begin{array}{l} 20 \times 7 = 140, \text{ write 40, carry 1 hundred.} \\ 20 \times 50 = 1000, \text{ add carried 100 to get 1100, write 11.} \end{array}}$

$\leftarrow$ Add.

The zero in 1140 is usually omitted to save time.

Try these:

(a)   64
      × 37

(b)   327
      × 145

(c)   342
      × 102

Work each problem as shown above. Use the shortcut method if possible. Check your answers in **33**.

33 (a) **Guess:** $60 \times 40 = 2400$

$$
\begin{array}{r}
\overset{1}{\underset{2}{6}}4 \\
\times \quad 37 \\
\hline
448 \\
1920 \\
\hline
2368
\end{array}
$$

| 7 × 4 = 28, write 8, carry 2 tens. |
| 7 × 60 = 420, add carried 20 to get 440, write 44. |
| 30 × 4 = 120, write 20, carry 1 hundred. |
| 30 × 60 = 1800, add carried 100 to get 1900, write 19. |

2368 ◄—— Add.

Check: $2400 \approx 2368$

(b) **Guess:** $300 \times 150 = 45{,}000$

$$
\begin{array}{r}
\overset{1}{}\overset{2}{} \\
\overset{1}{}\overset{3}{} \\
327 \\
\times \quad 145 \\
\hline
1635 \\
13080 \\
32700 \\
\hline
47415
\end{array}
$$

| 5 × 7 = 35, write 5, carry 3 tens. |
| 5 × 20 = 100, add carried 30 to get 130, write 3, carry 1 hundred. |
| 5 × 300 = 1500, add carried 100 to get 1600, write 16. |
| 40 × 7 = 280, write 80, carry 2 hundreds. |
| 40 × 20 = 800, add carried 200 to get 1000, write 0, carry 10 hundreds. |
| 40 × 300 = 12,000, add carried 1000 to get 13,000, write 13. |
| 100 × 327 = 32,700 |

Check: $45{,}000 \approx 47{,}415$

(c) **Guess:** $350 \times 100 = 35{,}000$

$$
\begin{array}{r}
342 \\
\times \quad 102 \\
\hline
684 \\
000 \\
34200 \\
\hline
34884
\end{array}
$$

684 ◄——— $2 \times 342 = 684$
000 ◄——— $0 \times 342 = 000$
34200 ◄——— $100 \times 342 = 34{,}200$

Check: $35{,}000 \approx 34{,}884$

Be very careful when there are zeros in the multiplier; it is very easy to misplace one of those zeros. Do not skip any steps and be sure to guess 'n check.

Go to **34** for a set of practice problems on multiplication of whole numbers.

Multiplication by a Multiple of 10

When either of the factors being multiplied ends in one or more zeros, follow these rules.

- To multiply by 10, attach a zero on the right.

    $86 \times 10 = 860$

- To multiply by 100, attach two zeros on the right.

    $7 \times 100 = 700$ $23 \times 100 = 2300$

- To multiply by 1000, attach three zeros on the right.

    $15 \times 1000 = 15,000$ $425 \times 1000 = 425,000$

- To multiply by any multiple of 10, multiply the nonzero parts first, and then attach the total number of right-end zeros.

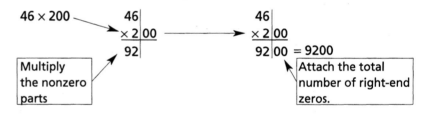

Multiply the nonzero parts

Attach the total number of right-end zeros.

Example: $2070 \times 4300 = ?$

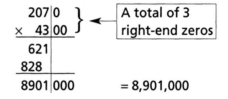

$$
\begin{array}{r}
207|0 \\
\times \quad 43|00 \\
\hline
621 \\
828 \\
\hline
8901|000
\end{array}
$$

A total of 3 right-end zeros

$= 8,901,000$

Try these problems for practice in multiplying numbers that end in zero.

1. $11 \times 10 =$ **2.** $100 \times 35 =$

3. $422 \times 10 =$ **4.** $501 \times 100 =$

5. $652 \times 100 =$ **6.** $723 \times 1000 =$

7. $403 \times 1000 =$ **8.** $43 \times 200 =$

9. $107 \times 2000 =$ **10.** $20,400 \times 100 =$

11. $4001 \times 300 =$ **12.** $2000 \times 3000 =$

13. $400 \times 30,200 =$ **14.** $600 \times 1200 =$

15. $5000 \times 21,000 =$

Check your answers in the Appendix.

34 *Problem Set 1-4:*   Multiplying Whole Numbers

A. Multiply:

1. 7	2. 9	3. 7	4. 7	5. 6	6. 9
6	8	8	9	8	6

7. 8	8. 6	9. 6	10. 9	11. 8	12. 8
9	7	8	7	7	6

B. Multiply:

1. 29	2. 67	3. 72	4. 27	5. 47
3	6	8	9	6

6. 88	7. 64	8. 37	9. 39	10. 42
9	5	7	4	7

11. 58	12. 87	13. 94	14. 49	15. 17
5	3	6	8	9

16. 23	17. 47	18. 53	19. 77	20. 36
7	6	8	4	9

21. 48	22. 35	23. 64	24. 72	25. 90
15	43	27	38	56

26. 41 72	**27.** 86 83	**28.** 18 65	**29.** 34 57	**30.** 28 91
31. 66 25	**32.** 71 19	**33.** 59 75	**34.** 18 81	**35.** 29 32
36. 82 76	**37.** 78 49	**38.** 35 58	**39.** 94 95	**40.** 43 64

C. Multiply:

1. 305 123	**2.** 145 516	**3.** 3006 125	**4.** 481 203	**5.** 8043 37
6. 765 502	**7.** 809 47	**8.** 1107 98	**9.** 3706 102	**10.** 4210 304
11. 708 58	**12.** 6401 773	**13.** 684 45	**14.** 319 708	**15.** 2043 670
16. 354 88	**17.** 2008 198	**18.** 923 47	**19.** 563 107	**20.** 8745 583
21. 206 301	**22.** 600 210	**23.** 502 106	**24.** 236 125	**25.** 671 202
26. 340 260	**27.** 270 344	**28.** 3005 201	**29.** 1204 300	**30.** 3202 421
31. 3301 1004	**32.** 5500 2006	**33.** 7045 3000	**34.** 1250 3002	**35.** 4005 2006

D. Brain Boosters

1. A room has 26 square yards of floor space. If carpeting costs $23 per square yard, what would it cost to carpet the room?

2. How many hours are there in a normal 365-day year?

3. If you manage to save $23 per week, how much money will you have in a year (52 weeks)?

4. A portable TV can be bought on credit for $145 down and twelve payments of $95 each. What is its total cost?

5. How many chairs are in a theater with 32 rows of 26 seats each?

6. The Maharaja of Pourboy bought 16 Cadillacs, one for each of his wives, at a cost of $48,165 each. What was the total cost of his purchase?

7. Multiply 123,456,789 by 8 and add 9 to the product. What is the total?

8. Complete each of these:

(a) 37	(b) 101	(c) 271	(d) 1221	(e) 4649	(f) 7373
$\times 3$	$\times 11$	$\times 41$	$\times 91$	$\times 239$	$\times 1507$

E. Calculator Problems

1. $2381 \times 40{,}023 =$ **2.** $81{,}432 \times 43{,}000 =$

3. $6504 \times 22{,}680 =$ **4.** $11{,}701 \times 2004 =$

5. $45 \times 38 \times 324 \times 7 =$ **6.** $45 \times 45 \times 57 \times 21 =$

7. $23 \times 34 \times 45 \times 56 =$ **8.** $1{,}120{,}000 \times 3{,}450{,}000 =$

9. What is interesting about these three multiplications?

(a) 11,313	(b) 11,317	(c) 31,311
$\times 10{,}913$	$\times 10{,}917$	$\times 30{,}911$

10. Calculate:

$7 \times 7 =$	$7 \times 9 =$
$67 \times 67 =$	$77 \times 99 =$
$667 \times 667 =$	$777 \times 999 =$
$6667 \times 6667 =$	$7777 \times 9999 =$

Can you find the pattern in these multiplications?

The answers to these problems are in the Appendix. When you have had the practice you need, turn to **35** to study the division of whole numbers or return to the preview on page 1 and use it to determine where to go next.

5 Dividing Whole Numbers

35 Division is the reverse process for multiplication. It enables us to separate a given quantity into equal parts. The mathematical phrase $12 \div 3$ is read "twelve divided by three" and it asks us to separate a collection of 12 objects into 3 equal parts. The mathematical phrases

$$12 \div 3 \qquad 3\overline{)12} \qquad \frac{12}{3} \qquad 12/3$$

all represent division and are all read "twelve divided by three."

Perform this division:

$$12 \div 3 = \underline{\quad}$$

Go to **36** to continue.

36

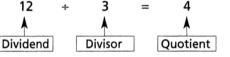

Divisor

Quotient

In this division problem, the number being divided (12) is called the **dividend**, the number used to divide (3) is called the **divisor**, and the result of the division (4) is called the **quotient**. The word "quotient" comes from a Latin word meaning "how many times."

One way to perform division is to reverse the multiplication process.

$$24 \div 4 = \square \text{ means that } 4 \times \square = 24$$

If the one-digit multiplication tables are firmly set in your memory, you will recognize immediately that $\square = 6$ in this problem.

Try these.

$35 \div 7 = \underline{\quad}$	$42 \div 6 = \underline{\quad}$
$28 \div 4 = \underline{\quad}$	$56 \div 7 = \underline{\quad}$
$45 \div 5 = \underline{\quad}$	$18 \div 3 = \underline{\quad}$
$70 \div 10 = \underline{\quad}$	$63 \div 9 = \underline{\quad}$
$30 \div 5 = \underline{\quad}$	$72 \div 8 = \underline{\quad}$

Check your answers in **37**.

37

$35 \div 7 = 5$	$42 \div 6 = 7$
$28 \div 4 = 7$	$56 \div 7 = 8$
$45 \div 5 = 9$	$18 \div 3 = 6$
$70 \div 10 = 7$	$63 \div 9 = 7$
$30 \div 5 = 6$	$72 \div 8 = 9$

You should be able to do all of these quickly by working backward from the one-digit multiplication table.

How do we divide dividends that are larger than 9×9 and therefore not in the multiplication table? Obviously, we need a better procedure. One way to learn how many times the divisor divides the dividend is to subtract it repeatedly. For example, in $12 \div 3$

12	9	6	3	3 is subtracted from 12 four
-3	-3	-3	-3	times, so that $12 \div 3 = 4$.
9	6	3	0	

Try it. Perform the division $138 \div 23$ using repeated subtraction. Check your answer in **38**.

Averages

Averages   The *average* of a set of numbers is the single number that best represents the whole set. One simple kind of average is the arithmetic average or arithmetic mean, defined as

$$\text{Arithmetic average} = \frac{\text{sum of measurements}}{\text{number of measurements}}$$

For example, the average weight of the five linebackers on our college football team is figured on page 48.

$$\text{Average weight} = \frac{215\ lb + 235\ lb + 224\ lb + 212\ lb + 239\ lb}{5}$$

$$= \frac{1125\ lb}{5} = 225\ lb$$

Try these problems for practice. Find the average of each of the following sets of numbers.

1. 27 and 85   **2.** 37, 26, and 45

3. 4, 12, 17, 3, 6, and 18   **4.** 1, 2, 3, 4, 5, 10, 10

5. 12, 14, 19, 23, 12, 46, 31, 24, 14, 15, 21

6. In a given week Maria worked in The Copy Place for the following hours each day: Monday, 3 hours; Tuesday, 3 hours; Wednesday, 7 hours; Thursday, 6 hours; Friday, 7 hours; and Saturday, 4 hours. What average amount of time does Maria work per day?

7. On four weekly quizzes in his history class, Willie scored 84, 74, 90, and 88 points. What is his average score?

8. A salesman sells widgets for the following amounts in successive weeks: $1647, $1710, $2205, $1349, and $1409. What is his average weekly sales?

The answers to these problems are in the Appendix.

38

$138 \div 23$	138	115	92	69	46	23
$\longrightarrow$	− 23	− 23	− 23	− 23	− 23	− 23
	115	92	69	46	23	0

23 may be subtracted from 138 six times, therefore $138 \div 23 = 6$.

We could also divide by simply guessing. For example, to find $245 \div 7$ by guessing, we might go through a mental conversation with ourselves something like this:

"7 into 245 goes how many times?"
  "Lots."
"How many? Pick a number."
  "Maybe 10."
"Let's try it: 7 × 10 = 70, so 10 is much too small. Try a larger number."
  "How about 50? Does 7 go into 245 about 50 times?"
"Well, 7 × 50 = 350, which is larger than 245. Try again."
  "I'm getting tired. Will 30 do it?"
"Well, 7 × 30 = 210. That is quite close to 245. Try something a little larger than 30."
  "31? 32? . . ."

Sooner or later the tired little guesser in your head will arrive at 35 and find that $7 \times 35 = 245$, and so

$245 \div 7 = 35$

With pure guessing even a simple problem could take all afternoon. We need a shortcut. The best division process combines one-digit multiplication, repeated subtraction, and educated guessing. For example, in the problem $96 \div 8$, start with a guess: the answer is about 10, since $8 \times 10 = 80$.

Step 1

Arrange the divisor and dividend horizontally.

Tens
column
Ones
column

$8\overline{)96}$

Can 8 be subtracted from 9? Yes, once.
Write 1 in the tens column, and place
a zero in the ones column. 10 is your
first guess at the quotient.

$$\begin{array}{r} 10 \\ 8\overline{)96} \end{array}$$

Step 2

Multiply $8 \times 10 = 80$.
Subtract $96 - 80 = 16$.

$$\begin{array}{r} 10 \\ 8\overline{)96} \\ -\ 80 \\ \hline 16 \end{array}$$

Step 3

Use 16 as the new dividend. Can 8 be
subtracted from 1? No. Write a zero
in the tens column above. Can 8 be
subtracted from 16? Yes, twice. Write
a 2 in the ones column above.

$$\begin{array}{r} 02 \\ 10 \\ 8\overline{)96} \\ -\ 80 \\ \hline 16 \end{array}$$

Step 4

Multiply $8 \times 2 = 16$.
Subtract $16 - 16 = 0$.

$$\begin{array}{r} 02 \\ 10 \\ 8\overline{)96} \\ -\ 80 \\ \hline 16 \\ -\ 16 \\ \hline 0 \end{array}$$

The quotient is the sum of the numbers in the answer space ($10 + 2 = 12$) so that $96 \div 8 = 12$. Always check your answer: $8 \times 12 = 96$.

Now you try one:

$112 \div 7 = \underline{\quad\quad}$

Work this problem using the method shown above. Then go to **39** .

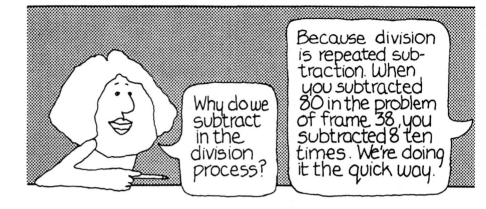

39   112 ÷ 7

Guess: $7 \times 10 = 70$ and $7 \times 20 = 140$, and so the answer is between 10 and 20.

Step 1

Can 7 be subtracted from 1? No. Write
a zero above the 1 in the hundreds column.
Can 7 be subtracted from 11? Yes, once.
Write a 1 in the tens column. Place a zero
in the ones column.

$$\begin{array}{r} 010 \\ 7\overline{)112} \end{array}$$

Step 2

Multiply $7 \times 10 = 70$ and subtract
$112 - 70 = 42$.

Step 3

Use 42 as the new dividend and
repeat the process: $7 \times 6 = 42$,
$42 - 42 = 0$.
Quotient $= 10 + 6 = 16$, remainder $= 0$.

$$\begin{array}{r} 6 \\ 010 \\ 7\overline{)112} \\ -\underline{70} \\ 42 \\ -\underline{42} \end{array}$$

Check: $7 \times 16 = 112$

When you get comfortable with this process, you can omit writing zeros and your work will look like this:

$$\begin{array}{r} 16 \\ 7\overline{)112} \\ \underline{7} \\ 42 \\ \underline{42} \\ 0 \end{array}$$

7 into 11 once, write 1.

11 – 7 = 4, bring down the 2.
7 into 42, 6 times, write 6.

7 × 6 = 42, 42 – 42 = 0.

Here are a few problems for practice in division.

(a) 976 ÷ 8   (b) 3174 ÷ 6   (c) 204 ÷ 6

Check your work in **40**.

Special Divisors

A few divisors require special attention. Remember that for any two numbers a and b, $a \div b = \square$ means that $b \times \square = a$. That is, $21 \div 3 = 7$ means $3 \times 7 = 21$.

1. If any number is divided by one, the quotient is the original number.

$$6 \div 1 = 6 \text{ and } \frac{6}{1} = 6 \text{ because } 6 \times 1 = 6$$

2. If any number is divided by itself, the quotient is one.

$$6 \div 6 = 1 \text{ and } \frac{6}{6} = 1 \text{ because } 6 \times 1 = 6$$

3. If zero is divided by any nonzero number, the quotient is zero.

$$0 \div 6 = 0 \text{ and } \frac{0}{6} = 0 \text{ because } 6 \times 0 = 0$$

4. If any number is divided by zero, the quotient is not defined in mathematics. If $6 \div 0 = \square$, then $0 \times \square = 6$, but 0 times any number equals zero. There is no number $\square$ that will make this equation true.

5. The fraction $\frac{0}{0}$ is never used because it can have any value whatever. If $\frac{0}{0} = \square$, then $0 = 0 \times \square$, and this equation is true for any value of $\square$.

---

40 (a) **Guess:** $8 \times 100 = 800$; the answer is something close to 100.

  Step 1 8 goes into 9 once, write 1.

  Step 2 $8 \times 1 = 8$, subtract $9 - 8 = 1$.

  Step 3 Bring down 7. 8 goes into 17 twice, write 2.

  Step 4 $8 \times 2 = 16$, subtract $17 - 16 = 1$.

  Step 5 Bring down 6. 8 goes into 16 twice, write 2.

  Step 6 $8 \times 2 = 16$, subtract $16 - 16 = 0$.

$$
\begin{array}{r}
122 \\
8\overline{)976} \\
-8 \\
\hline
17 \\
-16 \\
\hline
16 \\
-16 \\
\hline
0
\end{array}
$$

  Check: $8 \times 122 = 976$ Quotient = 122

(b) **Guess:** $6 \times 500 = 3000$; the answer is roughly 500.

  Step 1 6 into 3? No. 6 goes into 31 five times, write 5.

  Step 2 $6 \times 5 = 30$, subtract $31 - 30 = 1$.

  Step 3 Bring down 7. 6 goes into 17 twice, write 2.

  Step 4 $6 \times 2 = 12$, subtract $17 - 12 = 5$.

  Step 5 Bring down 4. 6 goes into 54 nine times, write 9.

  Step 6 $6 \times 9 = 54$, subtract $54 - 54 = 0$.

$$
\begin{array}{r}
529 \\
6\overline{)3174} \\
-30 \\
\hline
17 \\
-12 \\
\hline
54 \\
-54 \\
\hline
0
\end{array}
$$

Quotient = 529

  Check: $6 \times 529 = 3174$

(c) **Guess:** $6 \times 30 = 180$, and so the quotient is about 30.

  Step 1 6 into 2? No. 6 goes into 20 three times, write 3.

  Step 2 $6 \times 3 = 18$, subtract $20 - 18 = 2$.

  Step 3 Bring down 4. 6 goes into 24 four times, write 4.

  Step 4 $6 \times 4 = 24$, subtract $24 - 24 = 0$.

$$
\begin{array}{r}
34 \\
6\overline{)204} \\
-18 \\
\hline
24 \\
24 \\
\hline
0
\end{array}
$$

  Check: $6 \times 34 = 204$ Quotient = 34

Now try a problem using a two-digit divisor.

5084 ÷ 31 = _____

The procedure is the same as above. Check your answer in **41**.

Working with Measurement Numbers

Units In most practical work in arithmetic the numbers include units of some kind. A **denominate** number is one that describes the size of some quantity and includes a measurement unit.

$$\$3 = 3 \text{ dollars}$$
$$6 \text{ ft} = 6 \text{ feet or } 6 \times 1 \text{ foot}$$
$$10 \text{ lb} = 10 \text{ pounds or } 10 \times 1 \text{ lb}$$

Denominate numbers can be added, subtracted, multiplied, or divided like the whole numbers.

• To add or subtract denominate numbers, be certain that all numbers being combined have the same units. The answer will have these same units.

3 ft + 2 ft = 5 ft ———— All units must be the same.

```
3 + 2 = 5
Add the number
parts.
```

• To multiply or divide denominate numbers, work with the units and the number parts separately.

$$6 \text{ ft} \times 3 \text{ ft} = (6 \times 3)(1 \text{ ft} \times 1 \text{ ft})$$
$$= 18 \text{ sq ft}$$

$$50 \text{ miles} \div 2 \text{ hours} = (50 \div 2)(\text{miles} \div \text{hr})$$
$$= 25 \text{ miles per hr or 25 mph}$$

Try these problems for practice.

1. 34 ft + 12 ft – 19 ft =   **2.** 127 lb – 17 lb =
3. 14 yd + 21 ft – 24 in. = _____ ft   **4.** 45 sq ft + 21 sq ft =
5. 125 miles ÷ 5 hr =   **6.** 330 meters ÷ 10 sec =
7. 45 miles/hr × 2 hr =   **8.** 12 ft × 6 ft =
9. 3 in. × 12 in. × 5 in. =   **10.** 126 min + 1 hr = _____ min

Check your answers in the Appendix.

41   5084 ÷ 31

Guess: This is roughly the same as 500 ÷ 3 or about 200. The quotient will be about 200.

Step 1   31 into 5? No. 31 goes into 50 once, write 1.

Step 2   31 × 1 = 31, subtract 50 – 31 = 19.

Step 3   Bring down 8. 31 into 198? (That is about the same as 3 into 19.) Yes, 6 times, write 6.

Step 4   31 × 6 = 186, subtract 198 – 186 = 12.

Step 5   Bring down 4. 31 into 124? (That is about the same as 3 into 12.) Yes, 4 times, write 4.

Step 6   31 × 4 = 124, subtract 124 – 124 = 0.

```
        164
   31)5084
     - 31↓
       198
     - 186↓
       124
       124
         0
```

Quotient = 164

Check: 31 × 164 = 5084

Notice that in step 3 it is not at all obvious how many times 31 will go into 198. Again, you must make an educated guess and check your guess as you go along. If you guess 'n check on every problem you will always get the correct answer.

So far, we have looked only at division problems that "come out even." In these problems the remainder is zero. Obviously not all division problems are of this kind. What would you do with these?

(a) 59 ÷ 8 = ____   (b) 341 ÷ 43 = ____
(c) 7528 ÷ 37 = ____

Look in **42** for our answers.

42 (a) $\dfrac{7}{8\overline{)59}}$ The quotient is 7 with a remainder of 3.
         $-\ 56$
         $\overline{3}$

(b) $\dfrac{7}{43\overline{)341}}$ Your first guess would probably be that 43 goes into
       $-\ 301$ 341 eight times (try 4 into 34), but $43 \times 8 = 344$,
       $\overline{40}$ which is larger than 341. The quotient is therefore 7
                with a remainder of 40.

(c) $\dfrac{203}{37\overline{)7528}}$
       $-\ 74$
       $\overline{128}$ ← At this point notice that 37 cannot be subtracted from
       $-\ 111$ 12. Write a zero in the answer space and bring down
       $\overline{17}$ the 8. The quotient is 203, and the remainder is 17.

Now turn to **43** for a set of practice problems on division of whole
numbers.

43 *Problem Set 1-5:* Dividing Whole Numbers

A. Divide (Find the quotient and remainder if there is one):

1. $63 \div 7 =$ _____   **2.** $84 \div 7 =$ _____   **3.** $92 \div 8 =$ _____

4. $56 \div 8 =$ _____   **5.** $72 \div 0 =$ _____   **6.** $65 \div 5 =$ _____

7. $37 \div 5 =$ _____   **8.** $45 \div 9 =$ _____   **9.** $71 \div 7 =$ _____

10. $7 \div 1 =$ _____   **11.** $6 \div 6 =$ _____   **12.** $13 \div 0 =$ _____

13. $\dfrac{32}{4} =$ _____   **14.** $\dfrac{18}{3} =$ _____   **15.** $\dfrac{28}{7} =$ _____

16. $\dfrac{42}{6} =$ _____   **17.** $\dfrac{54}{9} =$ _____   **18.** $\dfrac{63}{7} =$ _____

B. Divide:

1. $245 \div 7 =$ _____   **2.** $369 \div 9 =$ _____   **3.** $167 \div 7 =$ ___

4. $126 \div 3 =$ _____   **5.** $228 \div 4 =$ _____   **6.** $232 \div 5 =$ ___

7. $310 \div 6 =$ _____   **8.** $360 \div 8 =$ _____   **9.** $337 \div 3 =$ ___

10. $\dfrac{132}{3} =$ _____   **11.** $\dfrac{147}{7} =$ _____   **12.** $\dfrac{216}{8} =$ _____

13. 7)364   **14.** 6)222   **15.** 8)704

16. 5)625   **17.** 4)201   **18.** 9)603

C. Divide:

1. 322 ÷ 14 = _____   **2.** 357 ÷ 17 = _____

3. 382 ÷ 19 = _____   **4.** 407 ÷ 13 = _____

5. 936 ÷ 24 = _____   **6.** 502 ÷ 10 = _____

7. 700 ÷ 28 = _____   **8.** 701 ÷ 36 = _____

9. 730 ÷ 81 = _____   **10.** $\frac{451}{11}$ = _____

11. $\frac{901}{17}$ = _____   **12.** $\frac{989}{23}$ = _____

13. 31)682   **14.** 43)507   **15.** 61)732

16. 12)408   **17.** 33)303   **18.** 13)928

D. Divide:

1. 2001 ÷ 21 = _____   **2.** 3328 ÷ 32 = _____

3. 2016 ÷ 21 = _____   **4.** 3536 ÷ 17 = _____

5. 1000 ÷ 7 = _____   **6.** 5029 ÷ 47 = _____

7. 2000 ÷ 9 = _____   **8.** 1881 ÷ 11 = _____

9. 2400 ÷ 75 = _____

10. 7)7000   **11.** 14)4275   **12.** 27)8405

13. 71)6005   **14.** 31)3105   **15.** 53)6307

16. 231)14,091   **17.** 411)42,020   **18.** 603)48,843

E. Brain Boosters

1. **Business**   The Wing Ding computer data entry specialist can type 75 words per minute. How long will it take him to key in a 7800-word document?

2. **Real Estate**   Sally earned $19,080 after taxes last year as a realtor. What monthly paycheck should she expect?

3. **Payroll**   The Pizza Palace has set aside $7280 for employee bonuses. If each of the 13 employees receives an equal bonus, how much does each employee receive?

4. Suppose your little Banzi sports car travels 528 miles on 11 gallons of gasoline. How many miles to the gallon are you getting?

5. Last week the People's Pinkey Company sold 213 pinkeys for a total of $10,863. What does one pinkey cost?

6. **Manufacturing**   If the Nutty Bolt Company produces 13,408 zinger bolts in a week and packages them 16 to a carton, how many cartons does it need?

7. The planet Pluto travels once around the sun in approximately 90,520 days. If one earth year is equal to 365 days, how many earth years is a Pluto year?

8. **Engineering**   The noise of an explosion travels 6210 meters through the air in 18 seconds. What is the speed of this sound? (Your answer should be in meters per second.)

F. Calculator Problems

1. 561,741 ÷ 123

2. 442,680 ÷ 408

3. 112,042 ÷ 371

4. 367,602 ÷ 591

5. 111,111 ÷ 111

6. 1,020,201 ÷ 10,101

7. 999,999 ÷ 6993

8. 964,348 ÷ 7777

The answers to these problems are in the Appendix. When you have had the practice you need, continue by going to the factoring of whole numbers in or by returning to the preview for this chapter on page 1.

6   Factors and Factoring

44   The symbols 6, VI, and ₦ᴋᴜ are all names for the number six. We can also write any number in terms of arithmetic operations involving other numbers. For example, $(4 + 2)$, $(7 - 1)$, (2×3), and $(18 \div 3)$ are also ways of writing the number six. It is particularly useful to be able to write any whole number as a product of other numbers. If we write

$6 = 2 \times 3$

Factors 2 and 3 are called the **factors** of 6. Of course we could write

$6 = 1 \times 6$

and see that 1 and 6 are also factors of 6, but this does not tell us anything new about the number 6. The factors of 6 are 1, 2, 3, and 6.

What are the factors of 12? (Choose an answer.)

(a) 2, 3, 4, and 6   Go to .
(b) 1, 2, 3, 4, 6, and 12   Go to **46**.
(c) 0, 1, 2, 3, 4, 6, and 12   Go to **47**.

 Not quite. Any two whole numbers whose product is 12 are factors of 12. It is easy to see that

$1 \times 12 = 12$

Therefore, 1 and 12 are factors of 12.

Return to **44** and choose a better answer.

46 Right you are.

$1 \times 12 = 12$ $2 \times 6 = 12$ $3 \times 4 = 12$

are all ways of writing 12 as the product of two numbers. Therefore, 1, 2, 3, 4, 6, and 12 are all factors of 12.

Evenly Divisible Any number is **evenly divisible** by its factors; that is, every factor divides the number with zero remainder. For example,

$$12 \div 1 = 12 \qquad 12 \div 4 = 3$$
$$12 \div 2 = 6 \qquad 12 \div 6 = 2$$
$$12 \div 3 = 4 \qquad 12 \div 12 = 1$$

List the factors of these numbers:

(a) 18   (b) 20   (c) 24   (d) 48

Check your answers in **48**.

 Not correct. Zero is never a factor of any number. There is no number $\square$ such that

$0 \times \square = 12$

The product of 0 and any number is always 0.

Return to **44** and choose a better answer.

48 (a) The factors of 18 are 1, 2, 3, 6, 9, and 18.

(b) The factors of 20 are 1, 2, 4, 5, 10, and 20.

(c) The factors of 24 are 1, 2, 3, 4, 6, 8, 12, and 24.

(d) The factors of 48 are 1, 2, 3, 4, 6, 8, 12, 16, 24, and 48.

For some numbers the only factors are 1 and the number itself. For example, the factors of 7 are 1 and 7 because

$1 \times 7 = 7$

Prime Numbers   There are no other numbers that divide 7 with remainder zero. Such numbers are known as **prime numbers.** A prime number is one for which there are no factors other than 1 and the prime number itself.

Here is a list of the first few prime numbers:

2	3	5	7	11
13	17	19	23	29
31	37	41	43	47

Notice that 1 is not listed. All prime numbers have two distinct, unequal factors: 1 and the number itself. The number 1 has only one factor—itself. The number 1 is not a prime.

There are twenty-five prime numbers less than 100, 168 less than 1000, and no limit to the total number. Mathematicians have tried for centuries to find a simple pattern that would enable them to write down the primes in order and predict if any given number is a prime. As yet no one has succeeded.

How then does one determine if some given number is a prime? There is no magic way to decide. You must divide the number by each whole number in order, starting with 2. If the division has no remainder, the original number is not a prime. Continue dividing until the quotient obtained is less than the divisor.

For example, is 53 a prime number? Try to work it out, and then turn to **49**.

49 To decide if 53 is a prime, we must perform a series of divisions:

$$4)\overline{53} \quad \begin{array}{c} 13 \\ \hline \end{array} \quad \text{remainder} = 1 \qquad 7)\overline{53} \quad \begin{array}{c} 7 \\ \hline \end{array} \quad \text{remainder} = 4$$

$$2)\overline{53} \quad \begin{array}{c} 26 \\ \hline \end{array} \quad \text{remainder} = 1 \qquad 5)\overline{53} \quad \begin{array}{c} 10 \\ \hline \end{array} \quad \text{remainder} = 3$$

$$3)\overline{53} \quad \begin{array}{c} 17 \\ \hline \end{array} \quad \text{remainder} = 2 \qquad 6)\overline{53} \quad \begin{array}{c} 8 \\ \hline \end{array} \quad \text{remainder} = 5$$

We can stop the search for a divisor here because dividing by 8 gives a quotient (6) less than the divisor (8). All divisors produce a nonzero remainder; therefore 53 is a prime.

LEARNING HELP ▷

It is really not necessary to test divide by 4 or 6. If the number is not evenly divisible by 2, it cannot be evenly divided by 4 or 6 because these are multiples of 2. In testing for primeness we need test divide only by primes. This will save time, but you must have the first few primes memorized.

For most of your work it will be sufficient if you remember the first eight or ten primes listed in **48**. A study card listing primes to be memorized is included in the back of this book. Use it.

Which of the following numbers are prime?

(a) 103	(b) 114	(c) 143	(d) 223
(e) 289	(f) 449	(g) 527	(h) 667

Test divide by primes as shown above, and then check your answers in **50**.

50 (a) $103 \div 2 = 51$ remainder 1 $103 \div 3 = 34$ remainder 1
$103 \div 5 = 20$ remainder 3 $103 \div 7 = 14$ remainder 5
$103 \div 11 = 9$ remainder 4

At this point there is no need to continue because the quotient (9) is less than the divisor (11). 103 is a prime.

(b) $114 \div 2 = 57$ remainder zero
Because 114 is evenly divisible by 2, it is not a prime.

Why isn't 1 a prime?

If we allowed 1 as a prime, then we could write any number as a product of primes in many ways. For example,
12 = 1×2×2×3 or
12 = 1×1×2×2×3 or
12 = 1×1×1×1×1×2×2×3
The fact that factoring into primes can only be done one way is important in mathematics.

(c) 143 ÷ 11 = 13 remainder zero
Because 143 is evenly divisible by 11 and 13, it is not a prime.

(d) 223 is a prime. Test divide it by 2, 3, 5, 7, and 11.

(e) 289 = 17 × 17; so 289 is not a prime.

(f) 449 is a prime. Test divide it by 2, 3, 5, 7, 11, 13, 17, and 19.

(g) 527 = 17 × 31; so 527 is not a prime.

(h) 667 = 23 × 29; so 667 is not a prime.

Prime Factors The **prime factors** of any numbers are those factors that are prime numbers.
The prime factors of 6 are 2 and 3. The prime factors of 21 are 3 and 7. The factors of 42 are 6 and 7, but the prime factors of 42 are 2, 3, and 7. The number 6 is not a prime factor; it is not a prime. To factor a number means to find its prime factors. Finding the prime factors of a number is a necessary skill if you want to learn how to add and subtract fractions.
What are the prime factors of 30?

(a) 5 × 6 = 30   Go to **51**.
(b) 2, 3, 5, 6, 10, 15   Go to **52**.
(c) 2, 3, 5   Go to **53**.

 You goofed on this one. The numbers 5 and 6 are factors of 30. Their product equals 30. But they are not both prime factors. Prime factors must be prime numbers.
Return to **50** and choose a different answer.

 These are some of the factors of 30, but not all of them are prime. Return to **50** and choose the set of prime factors.

53 Right! The prime factors of 30 are 2, 3, and 5.
Prime numbers are especially interesting because any number can be written as the product of primes in only one way. 6 = 2 × 3 and 6 = 3 × 2 count as only one way of writing 6 as a product of primes. The order is not important. The key idea is that 2 and 3 are the *only* primes whose product is 6.

9 = 3 × 3   This is a product of primes.

10 = 2 × 5   So is this.

11 = 11   An easy one, 11 is itself a prime. We would not
                   write 11 × 1 because 1 is not a prime . . . remember?

$12 = 2 \times 2 \times 3$

There are several ways to write 12 in terms of its factors ($12 = 2 \times 6$ or $12 = 3 \times 4$) but only one way with primes. Notice that the prime factor 2 must appear *twice*.

Write 60 as the product of its prime factors.

$60 =$ _____

Check your work in **54**.

The Sieve of Eratosthenes

More than 2000 years ago, Eratosthenes, a Greek geographer-astronomer, devised a way of locating primes that is still the most effective known. His procedure separates the primes out of the set of all whole numbers. Here is one version of what he did. First, arrange the whole numbers in six columns starting with 2 as shown. Second, circle the prime 2 and cross out all the multiples of 2; circle the next number (3) and cross out all multiples of 3; circle the next remaining number (5) and cross out all multiples of 5; and so on. The circled numbers remaining are the primes.

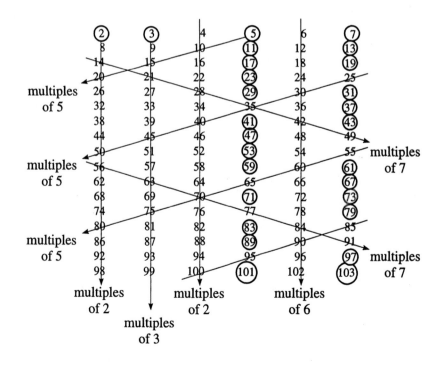

Mathematicians call this procedure a "sieve" because it is a way of sorting out or separating the primes from the other whole numbers.

Notice that all primes greater than 3 are either in the 5 or 7 column. They are either one less or one more than a multiple of 6. Pairs of primes separated by one integer (such as 5 and 7 or 11 and 13) are called *twin primes.* Can you find any other interesting patterns?

---

54 $60 = 2 \times 2 \times 3 \times 5$

Write the factors in any order you like, $2 \times 5 \times 3 \times 2$ or $5 \times 2 \times 3 \times 2$ or whatever. This is the *only* set of primes whose product is 60. It is this property of prime numbers that prompted the Greek mathematicians thirty centuries ago to call them "primes"—the "first" numbers from which the rest could be built.

Being able to write any number as a product of primes is a valuable skill. You will need this skill when you work with fractions in Chapter 2.

How can we rewrite a number in terms of its prime factors? Let's work through an example. Find the prime factors of 315. First, divide by the primes in order starting with 2.

$2 \lfloor \underline{315}$ 2 is not a factor because it does not divide the number evenly. Bring the number down and try the next prime.

$3 \lfloor \underline{315}$
$3 \lfloor \underline{105}$ 3 is a factor. Write it to the side ⟶ $\boxed{3}$ and divide by 3 again.
$3 \lfloor \underline{35}$

     3 is a factor again. Write it to the side ⟶ $\boxed{3}$ and divide by 3 again.

     3 does not divide into 35 evenly. Bring down 35 and try the next prime.

$5 \lfloor \underline{35}$
$\quad 7$ 5 is a factor. Write it to the side ⟶ $\boxed{5}$ and divide by 5 again.

     5 does not divide evenly into 7; try the next prime.

     7 is a prime factor; write it to the side. ⟶ $\boxed{7}$

     The number being factored is the product of the primes written on the right.

     $315 = 3 \times 3 \times 5 \times 7$

Finally, check this multiplication to be certain that you have missed no factors.

Find the prime factors of 693. Check your work in 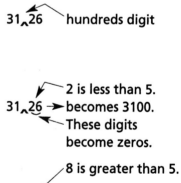 55.

Why are prime numbers called "prime"?

The Latin word <u>primus</u> means first in importance and the primes are the important main ingredients of numbers. Every number is either a prime or a product of primes.

Rounding Whole Numbers

Rounding To **round** a whole number means to find a number roughly equal to the given number but expressed less precisely.

Use the following procedure to round whole numbers.

Step 1 Place a mark ∧ to the **right** of the digit to which you are rounding.

Example Round 3126 to the nearest hundred.

31∧26 hundreds digit

Step 2 Look at the digit to the **right** of the mark.

- If this digit is *less than 5*, replace all digits to the right of the mark by zeros.

2 is less than 5.
31∧26 → becomes 3100.
These digits become zeros.

- If this digit is *equal to* or *greater than 5*, replace all digits to the right of the mark by zeros and add 1 to the digit on the left of the mark.

8 is greater than 5.
73∧81 → becomes 7400.

Practice rounding with these problems.

1. Round 47 to the nearest ten.

2. Round 675 to the nearest ten.

3. Round 255 to the nearest hundred. To the nearest ten.

4. Round 7,564 to the nearest hundred. To the nearest ten.

5. Round 26,398 to the nearest hundred. To the nearest thousand.

6. Round 5,309 to the nearest ten. To the nearest hundred.

7. Round 24,989 to the nearest ten. To the nearest thousand.

8. Round 65,555 to the nearest hundred. To the nearest thousand.

Check your answers in the Appendix.

The Factor Tree

A very helpful way to think about factors is in terms of a **factor tree.** For example, factor 1764.

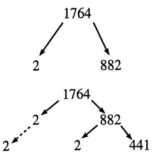

First, divide by the smallest prime, 2.
$1764 \div 2 = 882$.
Write down the 2 and the quotient 882.

Second, divide by 2 again.
$882 \div 2 = 441$.
On a new row write down both 2s and the quotient 441.

Third, when 2 will no longer divide the last quotient evenly, divide by the next largest prime, 3, . . . and so on. Stop when you find a prime quotient.

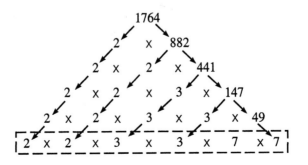

At each level of the tree the product of the horizontal numbers is equal to the original number to be factored. The last row gives the prime factors.

The Primes Less Than 100

2	3	5	7	11
13	17	19	23	29
31	37	41	43	47
53	59	61	67	71
73	79	83	89	97

---

55 Your work should look something like this:

$2\lfloor 693$ 2 does not divide 693 evenly.

$3\lfloor 693$
$3\lfloor 231$ $693 \div 3 = 231$ ⟶ ☐3
$3\lfloor 77$ $231 \div 3 = 77$ ⟶ ☐3
     3 does not divide 77 evenly.

$5\lfloor 77$ 5 does not divide 77 evenly.

$7\lfloor 77$ $77 \div 7 = 11$ ⟶ ☐7
   11   11 is a prime. ⟶ ☐11

$693 = 3 \times 3 \times 7 \times 11$ Check: $3 \times 3 \times 7 \times 11 = 9 \times 7 \times 11$
                                                 $= 63 \times 11 = 693$

Use this method to find the prime factors of these numbers:

(a) 570   (b) 792   (c) 945

Our work is in **56**.

56 (a) $2\lfloor 570$ $570 \div 2 = 285$ ⟶ ☐2
     $2\lfloor 285$ 2 is not a divisor of 285.

     $3\lfloor 285$ $285 \div 3 = 95$ ⟶ ☐3
     $3\lfloor 95$ 3 is not a divisor of 95.

     $5\lfloor 95$ $95 \div 5 = 19$ ⟶ ☐5
      19   19 is a prime ⟶ ☐19

     $570 = 2 \times 3 \times 5 \times 19$

(b) 2 |792   792 ÷ 2 = 396 ⟶ $\boxed{2}$
   2 |396   396 ÷ 2 = 198 ⟶ $\boxed{2}$
   2 |198   198 ÷ 2 = 99 ⟶ $\boxed{2}$
   2 |99   2 is not a divisor of 99.

   3 |99   99 ÷ 3 = 33 ⟶ $\boxed{3}$
   3 |33   33 ÷ 3 = 11 ⟶ $\boxed{3}$
     11   11 is a prime ⟶ $\boxed{11}$

792 = 2 × 2 × 2 × 3 × 3 × 11

(c) 2 |945   2 is not a divisor of 945.

   3 |945   945 ÷ 3 = 315 ⟶ $\boxed{3}$
   3 |315   315 ÷ 3 = 105 ⟶ $\boxed{3}$
   3 |105   105 ÷ 3 = 35 ⟶ $\boxed{3}$
   3 |35   3 is not a divisor of 35.

   5 |35   35 ÷ 5 = 7 ⟶ $\boxed{5}$
     7   7 is a prime ⟶ $\boxed{7}$

945 = 3 × 3 × 3 × 5 × 7

Divisibility   It is possible to tell at a glance, without actually dividing, if any number is evenly divisible by 2, 3, or 5. Knowing how to do so can save you a bit of work. Use these **divisibility rules.**

- Any number is evenly divisible by **2** if its ones digit is 2, 4, 6, 8, or 0.
  Examples: 12 is divisible by 2, it ends in 2.
  46 is divisible by 2, it ends in 6.
  7498 is divisible by 2, it ends in 8.
- Any number is evenly divisible by **3** if the sum of its digits is divisible by 3.
  Examples: 18 is divisible by 3 since 1 + 8 = 9 and 9 is divisible by 3.
  471 is divisible by 3 since 4 + 7 + 1 = 12 and 12 is divisible by 3.
  72,954 is divisible by 3 since 7 + 2 + 9 + 5 + 4 = 27 and 27 is divisible by 3.
  215 is *not* divisible by 3 since 2 + 1 + 5 = 8, and 8 is not divisible by 3.
- Any number is evenly divisible by **5** if its ones digit is 5 or 0.
  Examples: 25 is divisible by 5, it ends in 5.
  370 is divisible by 5, it ends in 0.
  73,495 is divisible by 5, it ends in 5.

There are divisibility rules for other numbers (shown in the following box, if you're interested), but the few above are all you really need to remember.

Use these rules to decide which of the following numbers are divisible by 2, 3, or 5. Do all work mentally. Check ✔ those evenly divisible by 2. Mark with an × those evenly divisible by 3. Circle those evenly divisible by 5.

16	23	27	39	45	111	132
330	335	372	453	498	785	921
1017	2111	73,908	123,456		4,271,305	

Turn to **57** to check your answers.

Divisibility Rules

Consider the following telephone number: 2615910. In a few seconds and without using pencil and paper, can you show that it is evenly divisible by 2, 3, 5, 6, 10, and 11 but not by 4, 7, 8, or 9? The trick is to use the following divisibility rules.

- **2**   Any number is divisible by **2** if its ones digit is 2, 4, 6, 8, or 0.
  Example: 14, 96, and 378 are all divisible by 2.

- **3**   Any number is evenly divisible by **3** if the sum of its digits is divisible by 3.
  Example: 672 is divisible by 3 since $6 + 7 + 2 = 15$ and 15 is divisible by 3.

- **4**   Any number is evenly divisible by **4** if the number formed by its two rightmost digits is divisible by 4.
  Example: 716 is divisible by 4 since 16 is divisible by 4.

- **5**   Any number is evenly divisible by **5** if its ones digit is 0 or 5.
  Example: 35, 90, and 1365 are all divisible by 5.

- **6**   Any number is divisible by **6** if it is divisible by both 2 and 3.
  Example: 822 is divisible by **6** since its ones digit is 2 and $8 + 2 + 2 = 12$ which is divisible by 3.

- **8**   Any number is divisible by **8** if its last three digits are divisible by 8.
  Example: 1160 is divisible by 8 since 160 is divisible by 8.

- **9**   Any number is divisible by **9** if the sum of its digits is divisible by 9.
  Example: 9243 is divisible by 9 since $9 + 2 + 4 + 3 = 18$ which is divisible by 9.

- **10** Any number whose ones digit is 0 is divisible by **10.**
  Example: 60, 210, and 19,830 are all divisible by 10.

There are no really simple rules for 7, 11, or 13. Here are the least complicated rules known.

- **7** Divide the number in question by 50. Add the quotient and remainder. The original number is divisible by 7 if the sum of the quotient and the remainder is divisible by 7.
  Example: 476 ÷ 50 = 9, remainder = 26.
  Add 9 + 26 = 35.
  35 is divisible by 7; therefore 476 is also divisible by 7.

- **11** Divide the number in question by 100. Add the quotient and remainder. The original number is divisible by 11 if the sum of the quotient and the remainder is divisible by 11.
  Example: 1562 ÷ 100 = 15, remainder = 62.
  Add 15 + 62 = 77.
  77 is divisible by 11; therefore 1562 is also divisible by 11.

- **13** Proceed as with 11, but divide by 40.
  Example: 1170 ÷ 40 = 29, remainder 10.
  Add 29 + 10 = 39.
  39 is divisible by 13; therefore 1170 is also divisible by 13.

---

57 ✔ means divisible by 2. × means divisible by 3. Circle means divisible by 5.

✔		×	×	×	×	✔×
16	23	27	39	⑤45	111	132

✔×		✔×	×	✔×		×
⑤330	⑤335	372	453	498	⑤785	921

×		✔×	✔×	
1017	2111	73,908	123,456	⑤4,271,305

Now you should be ready for some practice problems on factoring. Turn to **58** and continue.

Problem Set 1-6:   Factors and Factoring

A. List the prime factors of the following numbers:

1. 12_____	**2.** 16_____	**3.** 14_____
4. 18_____	**5.** 24_____	**6.** 20_____
7. 26_____	**8.** 31_____	**9.** 32_____
10. 36_____	**11.** 39_____	**12.** 42_____
13. 56_____	**14.** 81_____	**15.** 121_____

B. Write the following as products of primes:

1. 96 = _____	**2.** 84 = _____	**3.** 136 = _____
4. 170 = _____	**5.** 252 = _____	**6.** 256 = _____
7. 288 = _____	**8.** 390 = _____	**9.** 468 = _____
10. 546 = _____	**11.** 980 = _____	**12.** 1369 = _____
13. 1363 = _____	**14.** 1548 = _____	**15.** 3149 = _____

C. Circle the primes among the following numbers:

6	2	5	9	1	14	3	31	21	23
37	39	53	15	26	19	67	61	63	72
27	91	89	87	17					

D. Which of the following numbers are divisible by 2? Which by 3? Which by 5? (Do it in your head.)

9	12	4	231	45	144	17
261	1044	1390	72	81	102	2808
2088	8280	8802	11	111	1111	

E. Brain Boosters

1. A perfect number is one that is the sum of its divisors, not counting itself. For example, 6 = 1 + 2 + 3, and 1, 2, and 3 are divisors of 6, so 6 is a perfect number. Show that 28, 496, and 8128 are also perfect numbers.

2. The numbers 220 and 284 are called "amicable" or "friendly" numbers. It has been believed for hundreds of years that you can maintain a friendship by exchanging gifts, each related to one of these numbers. For example, in the Bible, Genesis 32:14, Jacob gave his brother Esau 220 goats and 220 sheep in an attempt to appease him and gain his friendship.

   a. What property of these numbers makes them special? (*Hint:* Add up the divisors of 284, not counting itself. Repeat for 220.)
   b. About 400 such pairs of friendly numbers are known, each made up of the parts of the other. Show that 2620 and 2924 are also a friendly pair.

3. Complete this magic square so that all rows, columns, and diagonals add to the same sum. What is special about the numbers in this square?

67	1	43
13		61
31	73	7

4. Consecutive numbers are whole numbers that differ by one. For example, 2 and 3 are consecutive; so are 115 and 116, or 734, 735, and 736. The product of three consecutive numbers is always divisible by what three numbers? Can you see why this is so?

5. Which of the following are primes? What are the prime factors of the others?

   | 1 | 11 | 111 | 1111 | 11,111 | 111,111 |

6. Fill in the missing digits marked with an asterisk (*):

```
    63*           *752          3*4
    *75           3*58          92*
  + 253           49*5        + *05
    *3*6        + 240*         1945
                  *5788
```

The answers to these problems are in the Appendix. When you have had the practice you need, continue by turning to **59** to study exponents and square roots or return to the preview on page 1.

7 Exponents and Square Roots

 When the same number appears many times as a factor, writing the product may become monotonous, tiring, and even inaccurate. It is easy, for example, to miscount the twos in

$$131{,}072 = 2 \times 2 \times 2 \times 2 \times 2 \times 2 \times 2 \times 2 \times 2 \times 2 \times 2 \times 2 \times 2 \times 2 \times 2 \times 2 \times 2$$

or the tens in

$$100{,}000{,}000{,}000 = 10 \times 10 \times 10 \times 10 \times 10 \times 10 \times 10 \times 10 \times 10 \times 10 \times 10$$

Products of this sort are usually written in a shorthand form as 2^{17} or 10^{11}. In this **exponential form** the raised number 17 indicates the number of times 2 is to be used as a factor. For example,

$$2 \times 2 = 2^2 \qquad \text{Product of } two \text{ factors of 2}$$
$$2 \times 2 \times 2 = 2^3 \qquad \text{Product of } three \text{ factors of 2}$$
$$\underbrace{2 \times 2 \times 2 \times 2}_{\text{Four 2s}} = 2^4 \qquad \text{Product of } four \text{ factors of 2}$$

Write $3 \times 3 \times 3 \times 3 \times 3$ in exponential form.

$$3 \times 3 \times 3 \times 3 \times 3 = \underline{\hspace{2cm}}$$

Check your answer in **60**.

---

60 $\underbrace{3 \times 3 \times 3 \times 3 \times 3}_{\text{Five factors of 3}} = 3^5$ $\qquad\qquad 3^5 \longleftarrow \boxed{\text{Exponent}}$

$\boxed{\text{Base}}$

Base

Exponent In this expression 3 is called the **base** and 5 is called the **exponent**. The exponent 5 tells you how many times the base 3 must be used as a factor in the product.

Multiply the factors in 4^3.

$$4^3 = \underline{\hspace{2cm}}$$

Pick an answer:

(a) 12   Go to **61**.

(b) 64   Go to **62**.

(c) 81   Go to **63**.

61 Your answer is incorrect; 4^3 is not equal to 12. The raised 3 in 4^3 tells you to multiply 4 by itself. Use the number 4 *three* times as a factor in a multiplication.

$$4^3 = \underbrace{4 \times 4 \times 4}$$

Means use three
factors of 4

Once you have set up the multiplication in this way it is easy to do it.

$$4 \times 4 \times 4 = (4 \times 4) \times 4 = 16 \times 4 = 64$$

Now, go on to **62**.

62 Excellent! In the exponential form 4^3, the exponent 3 tells you to multiply three factors of 4, that is,

$$4^3 = 4 \times 4 \times 4 = (4 \times 4) \times 4 = 16 \times 4 = 64$$

It is important that you be able to read exponential forms correctly.

2^2 is read "two to the second power" or "two squared"
2^3 is read "two to the third power" or "two cubed"
2^4 is read "two to the fourth power"
2^5 is read "two to the fifth power," and so on

Do the following problems to help get these concepts into your mental muscles. (Reading about a new concept may get it into, or at least through, your head, but only doing problems will make it part of you.)

(a) Write in exponential form:

$5 \times 5 \times 5 \times 5$ = _____ ; base = _____ exponent = _____

7×7 = _____ ; base = _____ exponent = _____

$10 \times 10 \times 10 \times 10 \times 10$ = _____ ; base = _____ exponent = _____

$3 \times 3 \times 3 \times 3 \times 3 \times 3 \times 3$ = _____ ; base = _____ exponent = _____

$9 \times 9 \times 9$ = _____ ; base = _____ exponent = _____

$1 \times 1 \times 1 \times 1$ = _____ ; base = _____ exponent = _____

(b) Write as a product of factors and multiply out:

2^6 = _____ = _____; base = _____ exponent = _____

10^7 = _____ = _____; base = _____ exponent = _____

3^4 = _____ = _____; base = _____ exponent = _____

5^2 = _____ = _____; base = _____ exponent = _____

6^3 = _____ = _____; base = _____ exponent = _____

4^5 = _____ = _____; base = _____ exponent = _____

12^3 = _____ = _____; base = _____ exponent = _____

1^5 = _____ = _____; base = _____ exponent = _____

The correct answers are in **64**. Go there when you finish these.

63 This answer is not correct.
Apparently you found the product $3 \times 3 \times 3 \times 3$.

$4^3 = 4 \times 4 \times 4$

Three factors of 4

The raised 3 tells you how many factors of 4 are to be multiplied together.

$4 \times 4 \times 4 = (4 \times 4) \times 4 = 16 \times 4 = ?$

Complete this and return to **60** to continue.

64 (a) $5 \times 5 \times 5 \times 5 = 5^4$; base = 5, exponent = 4
$7 \times 7 = 7^2$; base = 7, exponent = 2
$10 \times 10 \times 10 \times 10 \times 10 = 10^5$; base = 10, exponent = 5
$3 \times 3 \times 3 \times 3 \times 3 \times 3 \times 3 = 3^7$; base = 3, exponent = 7
$9 \times 9 \times 9 = 9^3$; base = 9, exponent = 3
$1 \times 1 \times 1 \times 1 = 1^4$; base = 1, exponent = 4

(b) $2^6 = 2 \times 2 \times 2 \times 2 \times 2 \times 2 = 64$; base = 2, exponent = 6
$10^7 = 10 \times 10 \times 10 \times 10 \times 10 \times 10 \times 10 = 10,000,000$; base = 10, exponent = 7
$3^4 = 3 \times 3 \times 3 \times 3 = 81$; base = 3, exponent = 4
$5^2 = 5 \times 5 = 25$; base = 5, exponent = 2

$6^3 = 6 \times 6 \times 6 = 216$; base = 6, exponent = 3
$4^5 = 4 \times 4 \times 4 \times 4 \times 4 = 1024$; base = 4, exponent = 5
$12^3 = 12 \times 12 \times 12 = 1728$; base = 12, exponent = 3
$1^5 = 1 \times 1 \times 1 \times 1 \times 1 = 1$; base = 1, exponent = 5

LEARNING HELP ▷ Any power of 1 is equal to 1.

$1^2 = 1 \times 1 = 1$
$1^3 = 1 \times 1 \times 1 = 1$
$1^4 = 1 \times 1 \times 1 \times 1 = 1$, and so on

LEARNING HELP ▷ When the base is ten, the product is easy to find.

$10^2 = 100$
$10^3 = 1000$
$10^4 = 10,000$
$10^5 = 100,000$, and so on

The exponent number for a power of 10 is always exactly equal to the number of zeros in the final product.

Look at that sequence of powers of ten again. Can you guess the value of 10^1? What about 10^0? Can you continue the pattern?

$10^1 = \underline{\qquad}$ $10^0 = \underline{\qquad}$

Try it, then turn to **65**.

65 In the pattern

$10^5 = 100,000$
$10^4 = 10,000$
$10^3 = 1,000$
$10^2 = 100$

each product on the right decreases by a factor of 10. Therefore, the next two lines must be

$10^1 = 10$
$10^0 = 1$

Of course, this is true for any base.

$2^1 = 2, 2^0 = 1$
$3^1 = 3, 3^0 = 1$
$4^1 = 4, 4^0 = 1$, and so on

If we factor 72 into its prime factors, we get

$72 = 2 \times 2 \times 2 \times 3 \times 3 = 2^3 \times 3^2$

If we factor the number 2592 into its prime factors, we find that

$2592 = 2 \times 2 \times 2 \times 2 \times 2 \times 3 \times 3 \times 3 \times 3$

Write this using exponents.

2592 = _____

Check your work in **66**.

66 $2592 = 2^5 \times 3^4$

Using exponents provides a simple and compact way to write any number as a product of its prime factors.

Find the following products by multiplying:

$2^4 \times 5^3 =$ _____ $3^4 \times 2^7 \times 1^6 =$ _____

$6^2 \times 7^3 =$ _____ $5^3 \times 8^2 \times 2^0 =$ _____

$2^3 \times 3^2 \times 5^4 =$ _____ $7^2 \times 9^1 \times 3^5 =$ _____

$2^2 \times 3^3 \times 4^4 =$ _____ $3^4 \times 5^2 \times 7^1 =$ _____

Go to **67** to check your answers.

67
$2^4 \times 5^3 = 2000$

$6^2 \times 7^3 = 12,348$

$2^3 \times 3^2 \times 5^4 = 45,000$

$2^2 \times 3^3 \times 4^4 = 27,648$

$3^4 \times 2^7 \times 1^6 = 10,368$

$5^3 \times 8^2 \times 2^0 = 8,000$

$7^2 \times 9^1 \times 3^5 = 107,163$

$3^4 \times 5^2 \times 7^1 = 14,175$

What is interesting about the following numbers?

1, 4, 9, 16, 25, 36, 49, 64, 81, 100

Do you recognize them? See **68**.

68   These numbers are the squares or second powers of the counting numbers.

$1^2 = 1$

$2^2 = 4$

$3^2 = 9$

$4^2 = 16$, and so on

Perfect Square   1, 4, 9, 16, 25, . . . are called **perfect squares**. If you have memorized the multiplication table for one-digit numbers, you will recognize them immediately. If you do not remember them, it will be helpful to you to memorize them. A study card for them is provided in the back of this book. Here are the first twenty perfect squares:

Perfect Squares

$1^2 = 1$	$6^2 = 36$	$11^2 = 121$	$16^2 = 256$
$2^2 = 4$	$7^2 = 49$	$12^2 = 144$	$17^2 = 289$
$3^2 = 9$	$8^2 = 64$	$13^2 = 169$	$18^2 = 324$
$4^2 = 16$	$9^2 = 81$	$14^2 = 196$	$19^2 = 361$
$5^2 = 25$	$10^2 = 100$	$15^2 = 225$	$20^2 = 400$

The number 3^2 is read "three squared." What is "square" about $3^2 = 9$? The name comes from an old Greek idea about the nature of numbers. Ancient Greek mathematicians called certain numbers "square numbers" or "perfect squares" because they could be represented by a square array of dots.

Square Root   The number of dots along the side of the square was called the "root" or origin of the square number. We call it the **square root**. For example, the square root of 16 is 4, since $4 \times 4 = 16$.

What is the square root of 64?

(a) 32   Go to **69**.
(b) 8   Go to **70**.

---

69   Sorry, you are incorrect. We cannot simply divide 64 in half to find its square root!

The square root of 64 is some number $\square$ such that $\square \times \square = 64$. For example, the square root of 25 is equal to 5 because $5 \times 5 = 25$. To "square" means to multiply by itself, and to find a square root means to find a number that when multiplied by itself gives the original number.

Now return to **68** and choose a better answer.

---

70   Right! The square root of 64 is equal to 8 because $8 \times 8 = 64$.

The sign $\sqrt{}$ is used to indicate the square root.
$\sqrt{16} = 4$ is read "square root of 16 equals 4."
$\sqrt{9} = 3$ is read "square root of 9 equals 3."
$\sqrt{169} = 13$ is read "square root of 169 equals 13."

Find these square roots:

$\sqrt{81} = $ _____ $\sqrt{361} = $ _____ $\sqrt{289} = $ _____

Try using the table in **68** if you do not recognize these. Continue in **71**.

Where did that funny little $\sqrt{}$ sign come from?

The word "root" is radix in Latin and about 1525 A.D. someone began to abbreviate it with the letter r in handwriting. Soon r led to r to √ to √.

71 $\sqrt{81} = 9$ Check: $9 \times 9 = 81$
$\sqrt{361} = 19$ Check: $19 \times 19 = 361$
$\sqrt{289} = 17$ Check: $17 \times 17 = 289$

Always check your answer as shown.

How do you find the square root of any whole number? There is no easy way to recognize or identify a perfect square and no quick and easy way to calculate the square root of a number. Furthermore, the square root of a number is usually not a whole number. Mathematicians have extended the Greek idea of square roots from the perfect squares to all numbers.

The surest and simplest way to find a square root is to use a calculator having a square root key: $\boxed{\sqrt{}}$ For example,

to find $\sqrt{15,129}$ 15,129 $\boxed{\sqrt{}}$ ⟶ $\boxed{\qquad 123 \qquad}$

In Chapter 3 we will extend the idea of square roots to decimal numbers and will show you how to find the square root of a number using a calculator.

Now, for a set of practice problems on exponents and square roots, turn to **72**.

72 *Problem Set 1-7:*   Exponents and Square Roots

A. Find the value of these:

1. $2^4 =$ ____	**2.** $3^2 =$ ____	**3.** $4^3 =$ ____	**4.** $5^3 =$ ____
5. $10^3 =$ ____	**6.** $7^2 =$ ____	**7.** $2^8 =$ ____	**8.** $6^2 =$ ____
9. $8^3 =$ ____	**10.** $3^4 =$ ____	**11.** $5^4 =$ ____	**12.** $10^5 =$ ____
13. $2^3 =$ ____	**14.** $3^5 =$ ____	**15.** $9^3 =$ ____	**16.** $7^0 =$ ____
17. $6^1 =$ ____	**18.** $1^4 =$ ____	**19.** $4^4 =$ ____	**20.** $2^5 =$ ____
21. $10^6 =$ ____	**22.** $7^3 =$ ____	**23.** $8^2 =$ ____	**24.** $6^4 =$ ____
25. $2^{10} =$ ____	**26.** $9^4 =$ ____	**27.** $6^3 =$ ____	**28.** $5^2 =$ ____
29. $3^3 =$ ____	**30.** $7^4 =$ ____	**31.** $9^0 =$ ____	**32.** $10^4 =$ ____
33. $4^2 =$ ____	**34.** $5^1 =$ ____	**35.** $8^4 =$ ____	**36.** $1^{14} =$ ____
37. $9^2 =$ ____	**38.** $4^5 =$ ____	**39.** $10^2 =$ ____	**40.** $6^5 =$ ____

B. Find the value of these:

1. $14^2 =$ ____

2. $21^2 =$ ____

3. $15^3 =$ ____

4. $25^3 =$ ____

5. $16^2 =$ ____

6. $55^2 =$ ____

7. $61^2 =$ ____

8. $40^3 =$ ____

9. $100^3 =$ ____

10. $2^2 \times 3^3 =$ ____

11. $2^6 \times 3^2 =$ ____

12. $3^2 \times 5^3 =$ ____

13. $2^1 \times 3^4 \times 7^2 =$ ____

14. $3^2 \times 7^2 \times 11^1 =$ ____

15. $2^0 \times 3^4 \times 5^2 =$ ___

16. $2^3 \times 7^3 =$ ____

17. $9^2 \times 2^4 =$ ____

18. $6^2 \times 5^2 \times 3^3 =$ ___

19. $2^1 \times 10^3 =$ ____

20. $3^2 \times 10^4 =$ ____

21. $2^{10} \times 3^2 =$ ____

C. Calculate these square roots:

1. $\sqrt{81} =$ ____

2. $\sqrt{144} =$ ____

3. $\sqrt{16} =$ ____

4. $\sqrt{25} =$ ____

5. $\sqrt{36} =$ ____

6. $\sqrt{100} =$ ___

7. $\sqrt{49} =$ ____

8. $\sqrt{324} =$ ____

9. $\sqrt{1} =$ ____

10. $\sqrt{121} =$ ____

11. $\sqrt{64} =$ ____

12. $\sqrt{9} =$ ____

13. $\sqrt{225} =$ ____

14. $\sqrt{4} =$ ____

15. $\sqrt{400} =$ ___

16. $\sqrt{256} =$ ____

D. Brain Boosters

1. Show that the following interesting equations are correct. (Can you see why they are interesting?)

$2^5 \times 9^2 = 2592$

$3^4 \times 425 = 34{,}425$

$12^2 + 33^2 = 1233$

$31^2 \times 325 = 312{,}325$

$88^2 + 33^2 = 8833$

$87^2 - 78^2 = 41^2 - 14^2$

$1^3 + 5^3 + 3^3 = 153$

$75^2 - 57^2 = 51^2 - 15^2$

$3^3 + 7^3 + 1^3 = 371$

$9^4 + 4^4 + 7^4 + 4^4 = 9474$

$4^3 + 0^3 + 7^3 = 407$

$3^3 + 4^4 + 3^3 + 5^5 = 3435$

$1^1 + 3^2 + 5^3 = 135$

$1^1 + 7^2 + 5^3 = 175$

$4^2 + 3^3 = 43$

$6^2 + 3^3 = 63$

$(4 + 9 + 1 + 3)^3 = 4913$ (add the four numbers in the parentheses, and then cube their sum)

2. Show that this set of equations is true by finding the value of each side.

$2 + 3 + 10 + 11 = 1 + 5 + 8 + 12$
$2^2 + 3^2 + 10^2 + 11^2 = 1^2 + 5^2 + 8^2 + 12^2$
$2^3 + 3^3 + 10^3 + 11^3 = 1^3 + 5^3 + 8^3 + 12^3$

3. Check the following claims for the first few powers.

(a) Every power of 5 ends in 5.
$5^2 = $ _____ $5^3 = $ _____ $5^4 = $ _____

(b) Every power of 25 ends in 25.
$25^2 = $ _____ $25^3 = $ _____ $25^4 = $ _____

(c) Every power of 625 ends in 625.
$625^2 = $ _____ $625^3 = $ _____ $625^4 = $ _____

(d) Every power of 376 ends in 376.
$376^2 = $ _____ $376^3 = $ _____ $376^4 = $ _____

The answers to these problems are in the Appendix. When you have had the practice you need, turn to **73** for a self-test covering the concepts learned in Chapter 1.

73 **CHAPTER 1 SELF-TEST**

1. $37 + 46 = $ _____

2. $4372 + 849 = $ _____

3. $81 + 47 + 36 = $ _____

4. $43 - 17 = $ _____

5. $702 - 416 = $ _____

6. $6001 - 3973 = $ _____

7. $8300 - 4605 = $ _____

8. $43 \times 36 = $ _____

9. $237 \times 204 = $ _____

10. $406 \times 137 = $ _____

11. $2153 \times 302 = $ _____

12. $1081 \div 47 = $ _____

13. $27\overline{)21{,}687}$ = _____

14. $2008 \div 4$ = _____

15. $\dfrac{26 \times 48}{39}$ = _____

16. Factor: 408 = _____

17. Factor: 9438 = _____

18. Factor: 1000 = _____

19. Factor: 3264 = _____

20. 3^4 = _____

21. $3^0 \times 4^2 \times 5^3$ = _____

22. $2^1 \times 3^4 \times 7^2$ = _____

23. 123^2 = _____

24. $\sqrt{225}$ = _____

25. $\sqrt{324}$ = _____

The answers to these problems are in the Appendix.

2 Fractions

PREVIEW 2

2. Multiply and divide fractions.

(a) $1\frac{1}{3} \times 2\frac{3}{5} =$ _____   98   19

(b) $\left(2\frac{2}{3}\right)^2 =$ _____   98   19

(c) $3\frac{1}{2} \div 1\frac{3}{4} =$ _____   105   26

(d) $6 \div 2\frac{1}{3} =$ _____   105   26

(e) Divide 6 by $2\frac{2}{5}$. _____   105   26

3. Add and subtract fractions.

(a) $4\frac{2}{3} + 1\frac{3}{4} =$ _____   114   34

(b) $5\frac{1}{8} - 3\frac{1}{3} =$ _____   114   34

(c) $9 - 1\frac{3}{8} =$ _____   114   34

4. Do word problems involving fractions.

(a) What fraction of $\frac{2}{3}$ is $1\frac{3}{4}$? _____   129   51

(b) Find $2\frac{1}{3}$ of $1\frac{7}{8}$. _____   129   51

(c) If 7 apples cost 91¢, what will 4   129   51
apples cost? _____

(d) Find a number such that $\frac{7}{8}$ of it is $2\frac{1}{2}$. _____   129   51

If you are certain you can work all of these problems correctly, turn to page 141 for a self-test. If you want help with any of these objectives or if you cannot work one of the preview problems, turn to the page indicated. Super-students (those who want to be certain they learn all of this) will join us in frame **1**.

2 FRACTIONS

By permission of Johnny Hart and Field Enterprises Inc.

1 Renaming Fractions

1 Measuring sticks require numbers as labels to identify their marks, and the space between marks can be specified using counting numbers or whole numbers: 1 inch, 2 inches, 5 feet, 12 cm, and so on. Fractions enable us to label some of the points between counting numbers. The word **fraction**

comes from the Latin *fractus* meaning "to break." We use fractions to describe subdivisions of the standard measurement units for length, time, money, or whatever we choose to measure.

Consider the rectangular area below. What happens when we break it into equal parts?

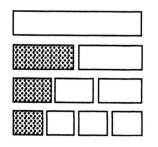

2 parts, each one-half of the whole

3 parts, each one-third of the whole

4 parts, each one-fourth of the whole

Divide the area below into fifths by drawing vertical lines.

Try it, then go to **2**.

2  Notice that the five parts or "fifths" are equal in area.

A fraction is normally written as the division of two whole numbers:

$$\frac{2}{3}, \frac{3}{4}, \text{ or } \frac{26}{12}$$

Each of the five equal areas above would be "one-fifth" or ⅕ of the entire area.

 $\frac{1}{5} = \frac{1 \text{ shaded part}}{5 \text{ parts total}}$

How would you label the shaded portion of the area below?

Continue in **3**.

3 ▨▨▨☐☐ $\dfrac{3}{5} = \dfrac{3 \text{ shaded parts}}{5 \text{ parts total}}$

The fraction ⅗ implies an area equal to three of the original portions.

$$\frac{3}{5} = 3 \times \left(\frac{1}{5}\right)$$

There are three equal parts and the name of each part is ⅕ or one-fifth.

In the collection of letters below, what fraction are **Hs**? (*Hint:* Count the total number of letters and decide what portion are **Hs**.)

H   H   H   H   S   S   T

Check your answer in **4**.

4 Fraction of **Hs** = $\dfrac{\text{number of } \mathbf{Hs}}{\text{total number of letters}} = \dfrac{4}{7}$

The answer, 4/7, is read "four-sevenths." The fraction of **Ss** is 2/7, and the fraction of **Ts** is 1/7.

Numerator
Denominator The two numbers that form a fraction are given special names to simplify talking about them. In the fraction ⅗ the upper number (3) is called the **numerator** from the Latin *numero* meaning number. It is a count of the number of parts. The lower number (5) is called the **denominator** from the Latin *nomen* or *name*. It tells us the name of the part being counted.

$\dfrac{3}{5}$ ← Numerator, the number of parts
    ← Denominator, the name of the part ("fifths," in this case)

A notebook costs $6 and I have $5. What fraction of its cost do I have? Write the answer as a fraction.

_____; numerator = ____, denominator = ____

Check your answer in **5**.

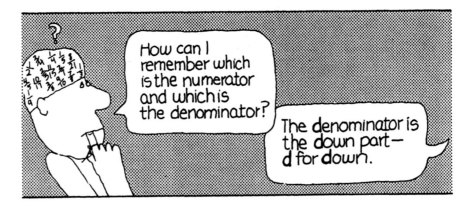

How can I remember which is the numerator and which is the denominator?

The denominator is the down part— d for down.

5 $$$$ $5 is $\dfrac{5}{6}$ of the total cost.

      5   Numerator = 5, denominator = 6.

Complete these sentences by writing in the correct fraction.

(a) If we divide a length into eight equal parts, each part will be _____ of the total length.

(b) Then three of these parts will represent _____ of the total length.

(c) Eight of these parts will be _____ of the total length.

(d) Ten of these parts will be _____ of the total length.

Check your answers in **6**.

6 (a) $\dfrac{1}{8}$ (b) $\dfrac{3}{8}$ (c) $\dfrac{8}{8}$ (d) $\dfrac{10}{8}$

Proper Fraction The original length is used as a standard, and any other length (smaller or larger) can be expressed as a fraction of the original length. A **proper fraction** is a number less than 1, as you would suppose a fraction should be. It represents a quantity less than the standard. For example,

$$\frac{1}{2}, \frac{2}{3}, \text{ and } \frac{17}{20}$$

are all proper fractions. Notice that for a proper fraction, the numerator is less than the denominator (the top number is less than the bottom number).

Improper Fraction An **improper fraction** is a number greater than 1 and represents a quantity greater than the standard. If a standard length is 8 inches, a length of 11 inches would be ¹¹⁄₈ of the standard. Notice that for an improper fraction the numerator is greater than the denominator (top number greater than the bottom number).

Circle the proper fractions in the following list.

$$\frac{3}{2} \quad \frac{3}{4} \quad \frac{7}{8} \quad \frac{5}{4} \quad \frac{15}{12} \quad \frac{1}{16} \quad \frac{35}{32} \quad \frac{7}{50} \quad \frac{65}{64} \quad \frac{105}{100}$$

Go to **7** when you have finished.

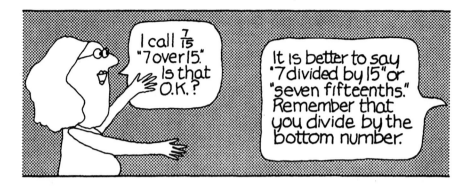

7 You should have circled the following proper fractions: ¾, ⅞, ¹⁄₁₆, ⁷⁄₅₀. All these are numbers less than 1. In each the numerator is less than the denominator.

The improper fraction ⅞ can be shown graphically as follows:

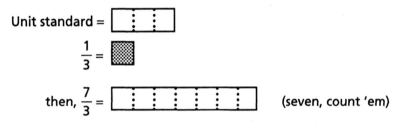

Unit standard =

$$\frac{1}{3} =$$

then, $\frac{7}{3} =$ (seven, count 'em)

We can rename this number by regrouping.

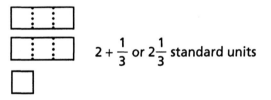

$2 + \dfrac{1}{3}$ or $2\dfrac{1}{3}$ standard units

Mixed Number A **mixed number** is an improper fraction written as the sum of a whole number and a proper fraction.

$$\frac{7}{3} = 2 + \frac{1}{3} \text{ or } 2\frac{1}{3}$$

We usually omit the plus sign and write $2 + ⅓$ as 2⅓, and read it as "two and one-third." The numbers 1½, 2⅗, and 16⅔ are all written as mixed numbers.

To write an improper fraction as a mixed number, divide numerator by denominator and form a new fraction as shown on the next page.

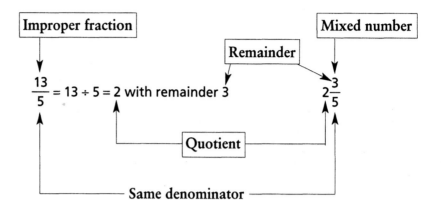

Now you try it. Rename ²³⁄₄ as a mixed number.

$$\frac{23}{4} = \underline{\quad\quad}$$

Follow the procedure shown above, then turn to **9**.

8   (a) $\dfrac{9}{5} = 1\dfrac{4}{5}$ (b) $\dfrac{13}{4} = 3\dfrac{1}{4}$ (c) $\dfrac{27}{8} = 3\dfrac{3}{8}$

(d) $\dfrac{31}{5} = 6\dfrac{1}{5}$ (e) $\dfrac{41}{12} = 3\dfrac{5}{12}$ (f) $\dfrac{17}{2} = 8\dfrac{1}{2}$

The reverse process, rewriting a mixed number as an improper fraction, is equally simple. Study the following example of 2⅗ being converted to an improper fraction.

LEARNING HELP ▷   • Work in a clockwise direction. First multiply $5 \times 2 = 10$.

$$2 \leftarrow \frac{3}{5}$$
$$\times$$

• Then add the numerator, $10 + 3 = 13$.

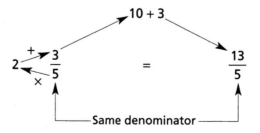

Here is how the same operation looks using a picture:

$2\frac{3}{5} = \left\{ \begin{array}{l} \\ \\ \\ \end{array} \right.$ $\begin{array}{c} \frac{5}{5} \\ \frac{5}{5} \\ \frac{5}{5} \\ \frac{3}{5} \end{array}$ $\frac{5}{5} + \frac{5}{5} + \frac{3}{5} = \frac{13}{5}$

(count 'em)

Now you try it. Rewrite these mixed numbers as improper fractions.

(a) $3\frac{1}{6}$ (b) $4\frac{3}{5}$ (c) $1\frac{1}{2}$

(d) $8\frac{2}{3}$ (e) $15\frac{3}{8}$ (f) $9\frac{3}{4}$

Check your answers in **10**.

9 $\frac{23}{4} = 23 \div 4 = 5$ with remainder $3 \to 5\frac{3}{4}$

If in doubt, check your work with a diagram like this:

• = • • • •
                    23   • • • • } 5 rows of 4
                                                • • • •
                                                • • • •
                                                • • • •
                                                • • •   3 remaining

Now try these for practice. Write each improper fraction as a mixed number.

(a) $\frac{9}{5}$ (b) $\frac{13}{4}$ (c) $\frac{27}{8}$ (d) $\frac{31}{5}$ (e) $\frac{41}{12}$ (f) $\frac{17}{2}$

The answers are in **8**.

10   (a) $3\frac{1}{6} = \frac{19}{6}$ (b) $4\frac{3}{5} = \frac{23}{5}$ (c) $1\frac{1}{2} = \frac{3}{2}$

(d) $8\frac{2}{3} = \frac{26}{3}$ (e) $15\frac{3}{8} = \frac{123}{8}$ (f) $9\frac{3}{4} = \frac{39}{4}$

Equivalent Fraction Two fractions are said to be **equivalent** if they are numerals or names for the same number. For example,

$$\frac{1}{2} = \frac{2}{4}$$

since both fractions represent the same portion of some standard amount.

$$\frac{1}{2} \quad \boxed{\blacksquare\ \ } \qquad = \frac{2}{4} \quad \boxed{\blacksquare\blacksquare\ \ }$$

There is a very large set of fractions equivalent to ½:

$$\frac{1}{2} = \frac{2}{4} = \frac{3}{6} = \frac{4}{8} = \frac{5}{10} = \cdots = \frac{46}{92} = \frac{61}{122} = \frac{1437}{2874} \cdots$$

Each fraction is a name for the same number, and we can use these fractions interchangeably.

To obtain a fraction equivalent to any given fraction, multiply the original (numerator and denominator) by the same nonzero number. For example,

$$\frac{1}{2} = \frac{1 \times 3}{2 \times 3} = \frac{3}{6} \qquad \frac{3}{6} \text{ is equivalent to } \frac{1}{2}.$$

$$\frac{2}{3} = \frac{2 \times 5}{3 \times 5} = \frac{10}{15} \qquad \frac{10}{15} \text{ is equivalent to } \frac{2}{3}.$$

Rename the fraction ¾ as an equivalent fraction with denominator 20.

$$\frac{3}{4} = \frac{?}{20}$$

Check your work in **11**.

11 $\dfrac{3}{4} = \dfrac{3 \times ?}{4 \times ?} = \dfrac{3 \times 5}{4 \times 5} = \dfrac{15}{20}$ $4 \times ? = 20$ so ? must be 5.

The number value of the fraction has not changed; we have simply renamed it.

Practice with these.

(a) $\dfrac{5}{6} = \dfrac{?}{42}$ (b) $\dfrac{7}{16} = \dfrac{?}{48}$

(c) $\dfrac{3}{7} = \dfrac{?}{56}$ (d) $1\dfrac{2}{3} = \dfrac{?}{12}$

Look in **13** for the answers.

12 $\dfrac{90}{105} = \dfrac{2 \times \not{3} \times 3 \times \not{5}}{\not{3} \times \not{5} \times 7} = \dfrac{2 \times 3}{7} = \dfrac{6}{7}$

Canceling This process of eliminating common factors is usually called **canceling**. When you cancel a factor, you *divide* both top and bottom of the fraction by that factor. In the cancellation above we actually divided top and bottom by 3×5 or 15.

$$\frac{90}{105} = \frac{90 \div 15}{105 \div 15} = \frac{6}{7}$$

CAUTION ⬎ • Cancel the same factor.

• Cancel only multiplying factors.

This is *illegal*: $\dfrac{2 \times \not{3} \times 5}{\not{3} \times \not{3} \times 7}$ Cancel only one 3. We divide the top by 3; so we must divide the bottom by 3.

This is *illegal*: $\dfrac{2 + \not{5}}{3 + \not{5}}$ 5 is not a multiplier. Canceling means *dividing* the top and bottom of the fraction by the same number. It is illegal to subtract a number from top and bottom.

Reduce the following fractions to lowest terms:

(a) $\dfrac{15}{84}$ (b) $\dfrac{21}{35}$ (c) $\dfrac{4}{12}$

(d) $\dfrac{154}{1078}$ (e) $\dfrac{256}{208}$ (f) $\dfrac{378}{405}$

The answers are in **14**.

13 (a) $\dfrac{5}{6} = \dfrac{5 \times \boxed{7}}{6 \times \boxed{7}} = \dfrac{35}{42}$ (b) $\dfrac{7}{16} = \dfrac{7 \times \boxed{3}}{16 \times \boxed{3}} = \dfrac{21}{48}$

(c) $\dfrac{3}{7} = \dfrac{3 \times \boxed{8}}{7 \times \boxed{8}} = \dfrac{24}{56}$ (d) $1\dfrac{2}{3} = \dfrac{5}{3} = \dfrac{5 \times \boxed{4}}{3 \times \boxed{4}} = \dfrac{20}{12}$

Lowest Terms Very often in working with fractions you will be asked to **reduce a fraction to lowest terms**. This means to replace the fraction with the simplest fraction in its set of equivalent fractions. For example, to reduce ¹⁵⁄₃₀ to its lowest terms, you would replace it with ½.

$$\frac{15}{30} = \frac{1 \times \boxed{15}}{2 \times \boxed{15}} = \frac{1}{2}$$

The two fractions ¹⁵⁄₃₀ and ½ are equivalent, and ½ is the simplest equivalent fraction to ¹⁵⁄₃₀ because its numerator (1) and denominator (2) are the smallest whole numbers of any in the set

$$\frac{1}{2}, \frac{2}{4}, \frac{3}{6}, \frac{4}{8}, \dots \frac{15}{30}, \dots$$

How can you find the simplest equivalent fraction? For example, how would you reduce ³⁰⁄₄₂ to lowest terms?

First, factor numerator and denominator.

$$\frac{30}{42} = \frac{2 \times 3 \times 5}{2 \times 3 \times 7}$$

Second, identify and eliminate common factors.

$$\frac{30}{42} = \frac{2 \times 3 \times 5}{2 \times 3 \times 7}$$

2 is a common factor, cancel the 2s: $\dfrac{30}{42} = \dfrac{\cancel{2} \times 3 \times 5}{\cancel{2} \times 3 \times 7}$

3 is a common factor, cancel the 3s: $\dfrac{30}{42} = \dfrac{\cancel{2} \times \cancel{3} \times 5}{\cancel{2} \times \cancel{3} \times 7}$

$$\frac{30}{42} = \frac{5}{7}$$

In effect, we have divided both top and bottom of the fraction by $2 \times 3 = 6$, the common factor.

$$\frac{30}{42} = \frac{30 \div 6}{42 \div 6} = \frac{5}{7}$$

Your turn. Reduce ⁹⁰⁄₁₀₅ to lowest terms. Look in **12** for the answer.

14

(a) $\dfrac{15}{84} = \dfrac{\cancel{3} \times 5}{2 \times 2 \times \cancel{3} \times 7} = \dfrac{5}{28}$ (b) $\dfrac{21}{35} = \dfrac{3 \times \cancel{7}}{5 \times \cancel{7}} = \dfrac{3}{5}$

(c) $\dfrac{4}{12} = \dfrac{\overset{1}{\cancel{4}}}{\cancel{4} \times 3} = \dfrac{1}{3}$ (d) $\dfrac{154}{1078} = \dfrac{\cancel{2} \times \cancel{7} \times \overset{1}{\cancel{11}}}{\cancel{2} \times \cancel{7} \times 7 \times \cancel{11}} = \dfrac{1}{7}$

(e) $\dfrac{256}{208} = \dfrac{\cancel{2} \times \cancel{2} \times \cancel{2} \times \cancel{2} \times 2 \times 2 \times 2 \times 2}{\cancel{2} \times \cancel{2} \times \cancel{2} \times \cancel{2} \times 13} = \dfrac{16}{13}$

(f) $\dfrac{378}{405} = \dfrac{2 \times \cancel{3} \times \cancel{3} \times \cancel{3} \times 7}{\cancel{3} \times \cancel{3} \times \cancel{3} \times 3 \times 5} = \dfrac{14}{15}$

CAUTION ▷   Remember that cancellation is division by a common factor.

Now check your understanding by reducing the fraction ⅚ to lowest terms. Check your answer in **15**.

15 $\dfrac{6}{3} = \dfrac{2 \times \cancel{3}}{1 \times \cancel{3}} = \dfrac{2}{1}$ (or simply 2)

LEARNING HELP Any whole number may be written as a fraction by using a denominator equal to 1.

$$3 = \frac{3}{1},\ 4 = \frac{4}{1},\text{ and so on}$$

The number 1 can be written as any fraction whose numerator and denominator are equal.

$$1 = \frac{2}{2} = \frac{3}{3} = \frac{4}{4} = \cdots = \frac{72}{72} = \frac{1257}{1257} \cdots \text{ and so on}$$

If you were offered your choice between ⅔ of a certain amount of money and ⅝ of the same amount of money, which would you choose? Which is the larger fraction, ⅔ or ⅝? Can you decide? Try. Renaming the fractions would help. The answer is in **16**.

16 To **compare** two fractions, rename each by changing them to equivalent fractions with the same denominator.

$$\frac{2}{3} = \frac{2 \times 8}{3 \times 8} = \frac{16}{24} \qquad \frac{5}{8} = \frac{5 \times 3}{8 \times 3} = \frac{15}{24}$$

Now compare the new fractions: ¹⁶⁄₂₄ is greater than ¹⁵⁄₂₄.

LEARNING HELP   1. The new denominator is the product of the original denominators ($24 = 8 \times 3$).

2. Once both fractions are written with the same denominator, the one with the larger numerator is the larger. (16 of the fractional parts is more than 15 of them.)

Which of the following pairs of fractions is larger?

(a) $\dfrac{3}{4}$ and $\dfrac{5}{7}$ (b) $\dfrac{7}{8}$ and $\dfrac{19}{21}$ (c) 3 and $\dfrac{40}{13}$

(d) $1\dfrac{7}{8}$ and $\dfrac{5}{3}$ (e) $2\dfrac{1}{4}$ and $\dfrac{11}{5}$ (f) $\dfrac{5}{16}$ and $\dfrac{11}{35}$

Check your answers in **17**.

17 (a) $\dfrac{3}{4} = \dfrac{21}{28}, \dfrac{5}{7} = \dfrac{20}{28}; \dfrac{21}{28}$ is larger than $\dfrac{20}{28}$ so $\dfrac{3}{4}$ is larger than $\dfrac{5}{7}$

(b) $\dfrac{7}{8} = \dfrac{147}{168}, \dfrac{19}{21} = \dfrac{152}{168}; \dfrac{152}{168}$ is larger than $\dfrac{147}{168}$ so $\dfrac{19}{21}$ is larger than $\dfrac{7}{8}$

(c) $3 = \dfrac{39}{13}; \dfrac{40}{13}$ is larger than $\dfrac{39}{13}$ so $\dfrac{40}{13}$ is larger than 3

(d) $1\dfrac{7}{8} = \dfrac{15}{8} = \dfrac{45}{24}, \dfrac{5}{3} = \dfrac{40}{24}; \dfrac{45}{24}$ is larger than $\dfrac{40}{24}$ so $1\dfrac{7}{8}$ is larger than $\dfrac{5}{3}$

(e) $2\dfrac{1}{4} = \dfrac{9}{4} = \dfrac{45}{20}, \dfrac{11}{5} = \dfrac{44}{20}; \dfrac{45}{20}$ is larger than $\dfrac{44}{20}$ so $2\dfrac{1}{4}$ is larger than $\dfrac{11}{5}$

(f) $\dfrac{5}{16} = \dfrac{175}{560}, \dfrac{11}{35} = \dfrac{176}{560}; \dfrac{176}{560}$ is larger than $\dfrac{175}{560}$ so $\dfrac{11}{35}$ is larger than $\dfrac{5}{16}$

Now turn to **18** for some practice renaming fractions.

18 *Problem Set 2-1:* Renaming Fractions

A. Write as an improper fraction:

1. $2\dfrac{1}{3}$ 2. $4\dfrac{2}{5}$ 3. $7\dfrac{1}{2}$ 4. $13\dfrac{3}{7}$ 5. $8\dfrac{3}{4}$

6. 4 7. $1\dfrac{2}{3}$ 8. $5\dfrac{5}{6}$ 9. $3\dfrac{7}{8}$ 10. $2\dfrac{3}{5}$

11. $16\frac{1}{10}$ 12. $70\frac{5}{9}$ 13. $12\frac{1}{40}$ 14. $15\frac{5}{11}$ 15. $37\frac{2}{3}$

B. Write as a mixed number:

1. $\frac{17}{2}$ 2. $\frac{23}{3}$ 3. $\frac{8}{5}$ 4. $\frac{19}{4}$ 5. $\frac{37}{6}$

6. $\frac{28}{3}$ 7. $\frac{37}{8}$ 8. $\frac{29}{7}$ 9. $\frac{34}{25}$ 10. $\frac{47}{9}$

11. $\frac{211}{4}$ 12. $\frac{170}{23}$ 13. $\frac{43}{10}$ 14. $\frac{125}{6}$ 15. $\frac{139}{15}$

C. Reduce to lowest terms:

1. $\frac{26}{30}$ 2. $\frac{12}{15}$ 3. $\frac{8}{10}$ 4. $\frac{27}{54}$ 5. $\frac{5}{40}$

6. $\frac{18}{45}$ 7. $\frac{7}{42}$ 8. $\frac{16}{18}$ 9. $\frac{9}{27}$ 10. $\frac{21}{56}$

11. $\frac{42}{120}$ 12. $\frac{54}{144}$ 13. $\frac{36}{216}$ 14. $\frac{280}{490}$ 15. $\frac{115}{207}$

D. Complete these:

1. $\frac{7}{8} = \frac{?}{16}$ 2. $\frac{3}{5} = \frac{?}{45}$ 3. $\frac{3}{4} = \frac{?}{12}$ 4. $2\frac{5}{12} = \frac{?}{60}$

5. $\frac{1}{9} = \frac{?}{63}$ 6. $1\frac{2}{7} = \frac{?}{35}$ 7. $\frac{5}{8} = \frac{?}{32}$ 8. $5\frac{3}{5} = \frac{?}{25}$

9. $\frac{1}{2} = \frac{?}{78}$ 10. $\frac{2}{3} = \frac{?}{51}$ 11. $8\frac{1}{4} = \frac{?}{44}$ 12. $5\frac{6}{7} = \frac{?}{14}$

13. $\frac{11}{12} = \frac{?}{72}$ 14. $3\frac{7}{10} = \frac{?}{50}$ 15. $9\frac{5}{9} = \frac{?}{54}$

E. **Brain Boosters**

1. The Gnu U. college track team is holding an indoor meet. The distance runners want to race 1 kilometer, a distance roughly equal to $^{62}/_{100}$ of a mile. Which is closer to 1 kilometer—6 laps on a track where each lap is $^{1}/_{11}$ of a mile or 5 laps on a track where each lap is $^{1}/_{8}$ of a mile?

2. Denny Dimwit wrote the following on his arithmetic exam:

$$\frac{1\cancel{6}}{\cancel{6}4} = \frac{1}{4} \qquad\qquad \frac{2\cancel{6}}{\cancel{6}5} = \frac{2}{5}$$

He claims that he has discovered that 6s always cancel. His teacher says that it is accidental that these illegal cancellations work. Write Denny a short explanation of what canceling really means.

3. A box of Sugar Glops breakfast cereal contains $\frac{7}{10}$ of a pound of cereal. A box of Astro Puffs contains 11 ounces of cereal (1 ounce = $\frac{1}{16}$ of a pound). The two cereals have the same food value (almost none!) and cost exactly the same amount. Which is the better buy?

4. **Nursing Assistant** Nurse Hypo was supposed to give his patient four $\frac{1}{2}$ grain No-Go pills. Instead, he gave her nine $\frac{1}{16}$ grain pills. Did he give her too much or too little?

5. Each of the following mixed numbers contains all nine nonzero digits, and they are all equal to the same whole number. What whole number are they equal to?

    (a) $91\dfrac{7524}{836}$ (b) $91\dfrac{5823}{647}$ (c) $94\dfrac{1578}{263}$

    (d) $96\dfrac{2148}{537}$ (e) $96\dfrac{1428}{357}$ (f) $96\dfrac{1752}{438}$

6. **Carpentry** Marta, an apprentice carpenter, measured the length of a two-by-four as $15\frac{5}{8}$ inches. Express this measurement in lowest terms.

7. **Sheet Metal Technology** Which is thicker, a $\frac{3}{16}$-inch sheet of metal or a $\frac{13}{64}$-inch fastener?

The answers to these problems are in the Appendix. When you have had the practice you need, either return to the preview test on page 83 or continue in **19** with the study of multiplication of fractions.

2 Multiplying Fractions

19 The simplest arithmetic operation with fractions is multiplication, and happily, it is easy to show graphically. The multiplication of a whole number and a fraction may be illustrated as follows.

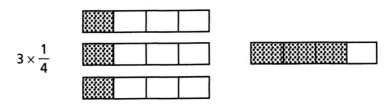

$$3 \times \frac{1}{4} = \frac{1}{4} + \frac{1}{4} + \frac{1}{4} = \frac{3}{4} \text{ (three segments each } \frac{1}{4} \text{ unit long)}$$

Any fraction such as ¾ can be thought of as a product: 3 × ¼. The product of two fractions can also be shown graphically.

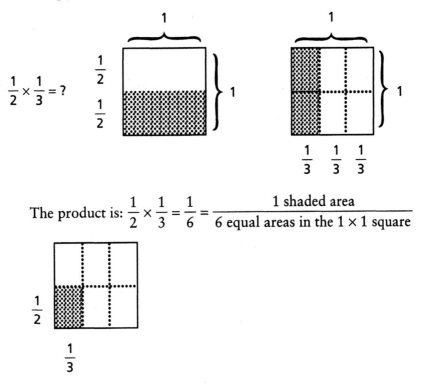

The product is: $\dfrac{1}{2} \times \dfrac{1}{3} = \dfrac{1}{6} = \dfrac{1 \text{ shaded area}}{6 \text{ equal areas in the } 1 \times 1 \text{ square}}$

LEARNING HELP ▷ Another way to solve this is ½ × ⅓ means ½ of ⅓.

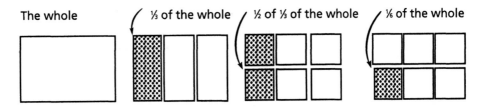

In general, we calculate this product as

$$\frac{1}{2} \times \frac{1}{3} = \frac{1 \times 1}{2 \times 3} = \frac{1}{6}$$

The product of two fractions is a fraction whose numerator is the product of their numerators and whose denominator is the product of their denominators.

Multiply $\frac{5}{6} \times \frac{2}{3}$, and then choose which answer below you think is correct.

(a) $\frac{5}{6} \times \frac{2}{3} = \frac{10}{18}$ Go to **20**.

(b) $\frac{5}{6} \times \frac{2}{3} = \frac{5}{9}$ Go to **21**.

(c) I don't know how to do it and I can't figure out how to draw the little boxes.   Go to **22**.

20 Right. $\frac{5}{6} \times \frac{2}{3} = \frac{5 \times 2}{6 \times 3} = \frac{10}{18}$

Now reduce this answer to lowest terms, and go to **21**.

21 Excellent. $\frac{5}{6} \times \frac{2}{3} = \frac{5 \times 2}{6 \times 3} = \frac{5 \times 2}{3 \times 2 \times 3} = \frac{5}{9}$

LEARNING HELP ▷ Always reduce your answer to lowest terms. In this problem you probably recognized that 6 was evenly divisible by 2 and did it like this:

$$\frac{5}{\overset{}{\underset{3}{6}}} \times \frac{\overset{1}{2}}{3} = \frac{5}{9}$$

You divided the top and the bottom of the fraction by 2. It will save you time and effort if you eliminate common factors (such as the 2 above) *before* you multiply.

Do you see that $3 \times \frac{1}{4}$ is really the same sort of problem?

$$3 = \frac{3}{1}, \text{ and so } 3 \times \frac{1}{4} = \frac{3}{1} \times \frac{1}{4} = \frac{3 \times 1}{1 \times 4} = \frac{3}{4}$$

Test your understanding with the following problems. (*Hint:* Change mixed numbers such as 1½ and 3⅚ into improper fractions; then multiply as usual.)

(a) $\frac{7}{8} \times \frac{2}{3} =$ _____   (b) $\frac{8}{12} \times \frac{3}{16} =$ _____

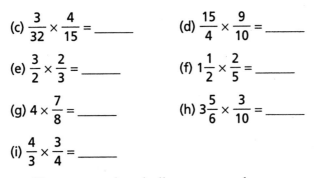

(c) $\dfrac{3}{32} \times \dfrac{4}{15} =$ _____

(d) $\dfrac{15}{4} \times \dfrac{9}{10} =$ _____

(e) $\dfrac{3}{2} \times \dfrac{2}{3} =$ _____

(f) $1\dfrac{1}{2} \times \dfrac{2}{5} =$ _____

(g) $4 \times \dfrac{7}{8} =$ _____

(h) $3\dfrac{5}{6} \times \dfrac{3}{10} =$ _____

(i) $\dfrac{4}{3} \times \dfrac{3}{4} =$ _____

Have you reduced all answers to lowest terms? The correct answers are in **23**.

22 Don't panic. You don't need to draw the little boxes to do the calculation. Try it this way:

$$\dfrac{5}{6} \times \dfrac{2}{3} = \dfrac{5 \times 2}{6 \times 3} \xleftarrow{\quad} \text{Multiply the numerators}$$
$$\xleftarrow{\quad} \text{Multiply the denominators}$$

Finish the calculation, and then return to **19** and choose an answer.

23 (a) $\dfrac{7}{8} \times \dfrac{2}{3} = \dfrac{7 \times \cancel{2}}{(4 \times \cancel{2}) \times 3} = \dfrac{7}{4 \times 3} = \dfrac{7}{12}$

CAUTION ▷ Eliminate common factors *before* you multiply. Your work will look like this when you learn to do these operations mentally:

$$\dfrac{7}{\underset{4}{\cancel{8}}} \times \dfrac{\overset{1}{\cancel{2}}}{3} = \dfrac{7}{12}$$

(b) $\dfrac{8}{12} \times \dfrac{3}{16} = \dfrac{\overset{1}{\cancel{8}} \times \overset{1}{\cancel{3}}}{(4 \times \cancel{3}) \times (\cancel{8} \times 2)} = \dfrac{1}{4 \times 2} = \dfrac{1}{8}$ or $\dfrac{\overset{1}{\cancel{8}}}{\underset{4}{\cancel{12}}} \times \dfrac{\overset{1}{\cancel{3}}}{\underset{2}{\cancel{16}}} = \dfrac{1}{8}$

(c) $\dfrac{\overset{3}{\cancel{3}}}{\underset{8}{\cancel{32}}} \times \dfrac{\overset{1}{\cancel{4}}}{\underset{5}{\cancel{15}}} = \dfrac{1}{40}$

(d) $\dfrac{\overset{3}{\cancel{15}}}{4} \times \dfrac{9}{\underset{2}{\cancel{10}}} = \dfrac{27}{8} = 3\dfrac{3}{8}$

(e) $\dfrac{\overset{1}{\cancel{3}}}{\underset{1}{\cancel{2}}} \times \dfrac{\overset{1}{\cancel{2}}}{\underset{1}{\cancel{3}}} = 1$

(f) $1\dfrac{1}{2} \times \dfrac{2}{5} = \dfrac{3}{\underset{1}{\cancel{2}}} \times \dfrac{\overset{1}{\cancel{2}}}{5} = \dfrac{3}{5}$

(g) $4 \times \dfrac{7}{8} = \dfrac{\overset{1}{\cancel{4}}}{1} \times \dfrac{7}{\underset{2}{\cancel{8}}} = \dfrac{7}{2} = 3\dfrac{1}{2}$

(h) $3\dfrac{5}{6} \times \dfrac{3}{10} = \dfrac{23}{\underset{2}{\cancel{6}}} \times \dfrac{\overset{1}{\cancel{3}}}{10} = \dfrac{23}{20} = 1\dfrac{3}{20}$

(i) $\dfrac{\overset{1}{\cancel{4}}}{\underset{1}{\cancel{3}}} \times \dfrac{\overset{1}{\cancel{3}}}{\underset{1}{\cancel{4}}} = 1$

Can you extend your new skills to solve these problems?

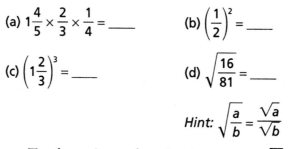

(a) $1\dfrac{4}{5} \times \dfrac{2}{3} \times \dfrac{1}{4} = $ ____

(b) $\left(\dfrac{1}{2}\right)^2 = $ ____

(c) $\left(1\dfrac{2}{3}\right)^3 = $ ____

(d) $\sqrt{\dfrac{16}{81}} = $ ____

Hint: $\sqrt{\dfrac{a}{b}} = \dfrac{\sqrt{a}}{\sqrt{b}}$

Try them. An explanation is waiting in **25**.

Using a Calculator with Fractions

Fractions can be entered directly on many calculators, and the results of calculations with fractions can be displayed as either fractions or decimals. If your calculator has a fraction key $\boxed{a\%}$, then you can use it to multiply, divide, add, and subtract fractions.

In this book we will explain the basic skills needed to work with fractions without a calculator. When you get to button-pushing you will understand the mathematical process better and make fewer mistakes. From time to time, we'll show how a calculator would do the work. For example,

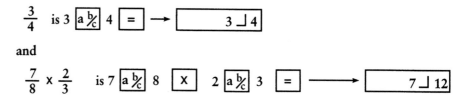

$\dfrac{3}{4}$ is 3 $\boxed{a\%}$ 4 $\boxed{=}$ → $\boxed{\qquad 3\lrcorner 4}$

and

$\dfrac{7}{8} \times \dfrac{2}{3}$ is 7 $\boxed{a\%}$ 8 $\boxed{\times}$ 2 $\boxed{a\%}$ 3 $\boxed{=}$ → $\boxed{\qquad 7\lrcorner 12}$

24   *Problem Set 2-2:*   Multiplying Fractions

A. Multiply and reduce the answer to lowest terms:

1. $\dfrac{1}{2} \times \dfrac{1}{4}$ **2.** $\dfrac{2}{3} \times \dfrac{1}{6}$ **3.** $\dfrac{2}{5} \times \dfrac{2}{3}$

4. $\dfrac{3}{8} \times \dfrac{1}{3}$ **5.** $\dfrac{4}{5} \times \dfrac{1}{6}$ **6.** $\dfrac{5}{3} \times \dfrac{1}{2}$

7. $6 \times \dfrac{1}{2}$ **8.** $\dfrac{5}{6} \times \dfrac{3}{5}$ **9.** $\dfrac{8}{9} \times 3$

10. $\dfrac{5}{16} \times \dfrac{8}{3}$ **11.** $\dfrac{11}{12} \times \dfrac{4}{15}$ **12.** $\dfrac{3}{7} \times \dfrac{3}{8}$

13. $\dfrac{8}{3} \times \dfrac{5}{12}$ **14.** $14 \times \dfrac{3}{4}$ **15.** $\dfrac{7}{8} \times \dfrac{13}{14}$

16. $\dfrac{5}{9} \times \dfrac{36}{25}$ **17.** $\dfrac{12}{8} \times \dfrac{15}{9}$ **18.** $\dfrac{32}{5} \times \dfrac{15}{16}$

19. $\dfrac{4}{7} \times \dfrac{49}{2}$ **20.** $\dfrac{16}{6} \times \dfrac{15}{28}$ **21.** $\dfrac{18}{5} \times \dfrac{10}{27}$

B. Multiply and reduce the answer to lowest terms:

1. $4\dfrac{1}{2} \times \dfrac{2}{3}$ **2.** $3\dfrac{1}{5} \times 1\dfrac{1}{4}$ **3.** $6 \times 1\dfrac{1}{3}$

4. $\dfrac{3}{8} \times 3\dfrac{1}{2}$ **5.** $2\dfrac{1}{6} \times 1\dfrac{1}{2}$ **6.** $7\dfrac{3}{4} \times 8$

7. $\dfrac{5}{7} \times 1\dfrac{7}{15}$ **8.** $1\dfrac{2}{9} \times \dfrac{3}{11}$ **9.** $4\dfrac{3}{5} \times 15$

10. $3\dfrac{3}{8} \times 1\dfrac{7}{9}$ **11.** $10\dfrac{5}{6} \times 3\dfrac{3}{10}$ **12.** $4\dfrac{5}{11} \times \dfrac{2}{7}$

13. $34 \times 2\dfrac{3}{17}$ **14.** $9\dfrac{7}{8} \times \dfrac{4}{5}$ **15.** $7\dfrac{9}{10} \times 1\dfrac{1}{4}$

16. $14 \times 3\dfrac{1}{3}$ **17.** $11\dfrac{6}{7} \times \dfrac{7}{8}$ **18.** $5\dfrac{1}{6} \times 2\dfrac{3}{5}$

19. $18 \times 1\dfrac{5}{27}$ **20.** $3\dfrac{1}{5} \times 1\dfrac{7}{8}$

C. Solve:

1. $\left(\dfrac{2}{3}\right)^2$ **2.** $\left(\dfrac{1}{2}\right)^4$ **3.** $\left(\dfrac{3}{5}\right)^3$

4. $\left(3\dfrac{1}{5}\right)^2$ **5.** $\left(4\dfrac{1}{2}\right)^3$ **6.** $\sqrt{\dfrac{9}{16}}$

7. $\sqrt{\dfrac{4}{49}}$ **8.** $\sqrt{\dfrac{16}{25}}$ **9.** $\sqrt{\dfrac{25}{64}}$

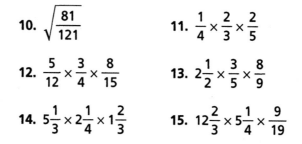

10. $\sqrt{\dfrac{81}{121}}$

11. $\dfrac{1}{4} \times \dfrac{2}{3} \times \dfrac{2}{5}$

12. $\dfrac{5}{12} \times \dfrac{3}{4} \times \dfrac{8}{15}$

13. $2\dfrac{1}{2} \times \dfrac{3}{5} \times \dfrac{8}{9}$

14. $5\dfrac{1}{3} \times 2\dfrac{1}{4} \times 1\dfrac{2}{3}$

15. $12\dfrac{2}{3} \times 5\dfrac{1}{4} \times \dfrac{9}{19}$

D. Brain Boosters

1. A jet plane cruises at 450 mph. How far does it travel in 3⅔ hours?

2. John ate ⅓ of a whole apple pie. Later Bert ate ¾ of the remainder. What part of the total pie did Bert eat?

3. The scale on a map is 1 centimeter equals 12½ kilometers. What actual distance is represented by a map distance of 8⅗ centimeters?

4. Ed spends ⅓ of his savings and gives ⅔ of the remainder to charity. He then has $12. How much money did he start with?

5. Show that the following calculation is correct:

$$\dfrac{18534}{9267} \times \dfrac{17469}{5823} = \dfrac{34182}{5697}$$

Notice that each fraction contains all nine nonzero digits.

6. **Real Estate**   What is the area in square miles of a farm 1⁵⁄₁₆ miles long by ⅔ mile wide?

7. **Nursing**   What is the total amount of medication in 6 pills each containing 2¾ milligrams?

8. **Construction**   Find the width of floor space covered by 38 boards each 3⅜ inch wide.

9. **Manufacturing**   How many pounds of elbow grease are contained in a barrel if a barrel holds 4½ gallons and a gallon of elbow grease weighs 7⅔ pounds?

10. **Photography**   A photograph must be reduced to four-fifths of its original size to fit the space available in a magazine. Find the length of the reduced photograph if the original was 6¾ inches long.

11. **Business**   Willie purchased 90 shares of IMD Corporation stock on WebStock.com at $4⅜ per share. What was the total cost?

The answers to these problems are in the Appendix. When you have had the practice you need, either return to the preview test on page 83 or continue in **26** with the study of the division of fractions.

25 (a) $1\dfrac{4}{5} \times \dfrac{2}{3} \times \dfrac{1}{4} = \dfrac{9}{5} \times \dfrac{2}{3} \times \dfrac{1}{4} = \dfrac{\overset{3}{\cancel{9}} \times \cancel{2} \times 1}{5 \times \cancel{3} \times \cancel{4}_{2}} = \dfrac{3}{10}$

Multiplication of three or more fractions involves nothing new. Be sure to change all mixed numbers to improper fractions before multiplying.

(b) $\left(\dfrac{1}{2}\right)^{2} = \dfrac{1}{2} \times \dfrac{1}{2} = \dfrac{1 \times 1}{2 \times 2} = \dfrac{1}{4}$ (Easy, right?)

(c) $\left(1\dfrac{2}{3}\right)^{3} = \left(\dfrac{5}{3}\right)^{3} = \dfrac{5}{3} \times \dfrac{5}{3} \times \dfrac{5}{3} = \dfrac{5 \times 5 \times 5}{3 \times 3 \times 3} = \dfrac{125}{27}$

Again, you must change any mixed numbers to improper fractions *before* you multiply.

(d) $\sqrt{\dfrac{16}{81}} = \dfrac{\sqrt{16}}{\sqrt{81}} = \dfrac{4}{9}$ Check: $\dfrac{4}{9} \times \dfrac{4}{9} = \dfrac{4 \times 4}{9 \times 9} = \dfrac{16}{81}$

Square roots are not difficult if both numerator and denominator are perfect squares. Otherwise you must use a calculator or a table of square roots and just divide decimals as shown in Chapter 3.

Now turn to **24** for a set of practice problems on multiplication.

---

3 Dividing Fractions

 Addition and multiplication are both reversible arithmetic operations. For example,

2×3 and 3×2 both equal 6
$4 + 5$ and $5 + 4$ both equal 9

The order in which you add addends or multiply factors is not important.

In subtraction and division this kind of exchange is *not* allowed, and because of this many people find these operations very troublesome.

$7 - 5 = 2$, but $5 - 7$ is *not* equal to 2 and is not even a counting number.
$8 \div 4 = 2$, but $4 \div 8$ is *not* equal to 2 and is not a whole number.

CAUTION ▷ In dividing fractions it is particularly important that you set up the process correctly.

Are these four numbers equal?

"8 divided by 4"   8)4   8 ÷ 4 $\frac{4}{8}$

Choose an answer:

(a) Yes   Go to **28**.
(b) No   Go to **29**.

27 The divisor is ½. The division 5 ÷ ½ is read "5 divided by ½," and it asks how many ½ unit lengths are included in a length of 5 units.

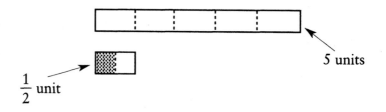

Division is defined in terms of multiplication.

8 ÷ 4 = □

asks that you find a number □ such that 8 = □ × 4. It is easy to see that □ = 2.

Division is defined in exactly the same way with fractions.

$5 \div \frac{1}{2} = \square$

asks that you find a number □ such that 5 = □ × ½.

Working backward from this multiplication and using the diagram above, find the answer to 5 ÷ ½. Hop ahead to **30** to continue.

28 Your answer is incorrect. Be very careful about this.

8 ÷ 4 is read "8 divided by 4" and written 4)8 or $\frac{8}{4}$

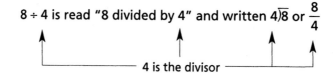

4 is the divisor

In the problem 8 ÷ 4 you are being asked to divide a set of 8 objects into sets of 4 objects. The divisor (4) is the denominator or bottom number of the fraction.

In the division 5 ÷ ½ which number is the divisor? Check your answer in **27**.

29 Right you are. "8 divided by 4" is written 8 ÷ 4. We can also write this as 4)8 or 8/4. In all of these the divisor is 4 and you are being asked to divide a set of 8 objects into sets of 4 objects.

In the division 5 ÷ ½ which number is the divisor? Check your answer in **27**.

30 $5 \div \dfrac{1}{2} = 10$

There are ten ½-unit lengths contained in the 5-unit length.

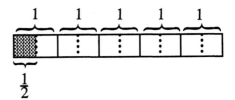

Using a drawing of this sort to solve a division problem is difficult and clumsy. We need a simple rule. Here it is:

To divide by a fraction, invert the divisor and multiply.

Example 1

Multiply

$$5 \div \frac{1}{2} = 5 \times \frac{2}{1} = 5 \times 2 = 10$$

The divisor ½ has been inverted.

To invert a fraction simply means to switch top and bottom. Inverting ⅔ gives 3/2. Inverting ⅕ gives 5/1. Inverting 7 gives 1/7.

Example 2

$$\frac{3}{5} \div \frac{2}{3} = \frac{3}{5} \times \frac{3}{2} = \frac{3 \times 3}{5 \times 2} = \frac{9}{10}$$

Using a calculator,

3 [a b/c] 5 [÷] 2 [a b/c] 3 [=] ⟶ [9⌐10]

We have converted a division problem that is difficult to picture into a simple multiplication. The final, and very important, step in every division is checking the answer.

If $\dfrac{3}{5} \div \dfrac{2}{3} = \dfrac{9}{10}$, then $\dfrac{3}{5} = \dfrac{2}{3} \times \dfrac{9}{10}$.

Check: $\dfrac{\cancel{2}^{1}}{\cancel{3}_{1}} \times \dfrac{\cancel{9}^{3}}{\cancel{10}_{5}} = \dfrac{3}{5}$

Why does this "invert and multiply" process work? The division ⅗ ÷ ⅔ = □ means that there is some number □ such that ⅗ = □ × ⅔. Multiply both sides of the last equation by ³⁄₂:

$$\left(\dfrac{3}{5}\right) \times \dfrac{3}{2} = \left(\square \times \dfrac{2}{3}\right) \times \dfrac{3}{2}$$

But we can do multiplication in any order we wish and we know that ⅔ × ³⁄₂ = ⁶⁄₆ = 1; so ⅗ × ³⁄₂ = □. Our answer, the unknown number we have labeled □, is simply the product of the dividend ⅗ and the inverted divisor ³⁄₂.

Try this one:

$$\dfrac{7}{8} \div \dfrac{3}{2} = \underline{\hspace{2cm}}$$

Solve it by inverting the divisor and multiplying. Check your answer in **31**.

31 $\dfrac{7}{8} \div \dfrac{3}{2} = \dfrac{7}{8} \times \dfrac{2}{3} = \dfrac{7 \times \cancel{2}^{1}}{\cancel{8}_{4} \times 3} = \dfrac{7}{12}$ Check: $\dfrac{\cancel{3}^{1}}{2} \times \dfrac{7}{\cancel{12}_{4}} = \dfrac{7}{8}$

The chief source of confusion in dividing fractions is deciding which fraction to invert. It will help if you:

LEARNING HELP ▷   • 1. First put every division problem in the form

(dividend) ÷ (divisor)

Then invert the divisor and multiply to obtain the quotient.

• 2. **Check** your answer by multiplying. The product

(divisor) × (quotient or answer)

should equal the dividend.

Here are a few problems to test your understanding.

(a) $\dfrac{2}{5} \div \dfrac{3}{8} = $ _____

(b) $\dfrac{7}{40} \div \dfrac{21}{25} = $ _____

(c) $3\dfrac{3}{4} \div \dfrac{5}{2} = $ _____

(d) $4\dfrac{1}{5} \div 1\dfrac{4}{10} = $ _____

(e) $3\dfrac{2}{3} \div 3 = $ _____

(f) Divide $\dfrac{3}{4}$ by $2\dfrac{5}{8}$. _____

(g) Divide 8 by $\dfrac{1}{2}$. _____

(h) Divide $1\dfrac{1}{4}$ by $1\dfrac{7}{8}$. _____

Work carefully, check each answer, and then turn to **32** for our worked solutions.

Why Do We Invert and Multiply to Divide Fractions?
The division 8 ÷ 4 can be written ⅛. Similarly, ½ ÷ ⅔ can be written

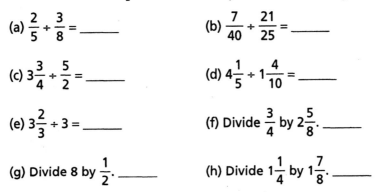

Notice the second fraction here (³⁄₂) is the original denominator inverted. Also, in the denominator $\frac{2}{3} \times \frac{3}{2} = \frac{2 \times 3}{3 \times 2} = \frac{6}{6} = 1$. Therefore, $\frac{1}{2} \div \frac{2}{3} = \frac{1}{2} \times \frac{3}{2}$. We have inverted the fraction ²⁄₃ and multiplied by it.

---

32 (a) $\frac{2}{5} \div \frac{3}{8} = \frac{2}{5} \times \frac{8}{3} = \frac{16}{15} = 1\frac{1}{15}$ Check: $\frac{\overset{1}{\cancel{3}}}{\cancel{8}} \times \frac{\overset{2}{\cancel{16}}}{\cancel{15}_{5}} = \frac{2}{5}$

Remember $\frac{3}{8} \times \frac{16}{15} = \frac{3 \times 16}{8 \times 15} = \frac{\cancel{3} \times 2 \times \cancel{8}}{\cancel{8} \times \cancel{3} \times 5} = \frac{2}{5}$

Using a calculator,

3 $\boxed{a\frac{b}{c}}$ 8 $\boxed{\times}$ 16 $\boxed{a\frac{b}{c}}$ 15 $\boxed{=}$ ⟶ $\boxed{\qquad 2\rfloor 5 \qquad}$

(b) $\frac{7}{40} \div \frac{21}{25} = \frac{7}{40} \times \frac{25}{21} = \frac{5}{24}$ Check: $\frac{\overset{7}{\cancel{21}}}{\cancel{25}_{5}} \times \frac{\overset{1}{\cancel{5}}}{\cancel{24}_{8}} = \frac{7}{40}$

(c) $3\frac{3}{4} \div \frac{5}{2} = \frac{15}{4} \div \frac{5}{2} = \frac{\overset{3}{\cancel{15}}}{\cancel{4}_{2}} \times \frac{\overset{1}{\cancel{2}}}{\cancel{5}_{1}} = \frac{3}{2} = 1\frac{1}{2}$ Check: $\frac{5}{2} \times \frac{3}{2} = \frac{15}{4} = 3\frac{3}{4}$

(d) $4\frac{1}{5} \div 1\frac{4}{10} = \frac{21}{5} \div \frac{14}{10} = \frac{\overset{3}{\cancel{21}}}{\cancel{5}_{1}} \times \frac{\overset{\overset{1}{\cancel{2}}}{\cancel{10}}}{\cancel{14}_{\cancel{7}}} = 3$

  Check: $1\frac{4}{10} \times 3 = \frac{14}{10} \times \frac{3}{1} = \frac{42}{10} = 4\frac{2}{10} = 4\frac{1}{5}$

(e) $3\frac{2}{3} \div 3 = \frac{11}{3} \div \frac{3}{1} = \frac{11}{3} \times \frac{1}{3} = \frac{11}{9} = 1\frac{2}{9}$

  Check: $3 \times \frac{11}{9} = \frac{3}{1} \times \frac{11}{9} = \frac{11}{3} = 3\frac{2}{3}$

(f) $\frac{3}{4} \div 2\frac{5}{8} = \frac{3}{4} \div \frac{21}{8} = \frac{\overset{1}{\cancel{3}}}{\cancel{4}_{1}} \times \frac{\overset{2}{\cancel{8}}}{\cancel{21}_{7}} = \frac{2}{7}$ Check: $2\frac{5}{8} \times \frac{2}{7} = \frac{21}{8} \times \frac{2}{7} = \frac{3}{4}$

(g) $8 \div \frac{1}{2} = \frac{8}{1} \times \frac{2}{1} = 16$ Check: $\frac{1}{2} \times 16 = 8$

(h) $1\frac{1}{4} \div 1\frac{7}{8} = \frac{5}{4} \div \frac{15}{8} = \frac{\overset{1}{\cancel{5}}}{\cancel{4}_{1}} \times \frac{\overset{2}{\cancel{8}}}{\cancel{15}_{3}} = \frac{2}{3}$

  Check: $1\frac{7}{8} \times \frac{2}{3} = \frac{15}{8} \times \frac{2}{3} = \frac{5}{4} = 1\frac{1}{4}$

Using a calculator,

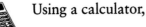 x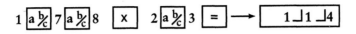

Turn to **33** for a set of practice problems on dividing fractions.

Ratio and Proportion

Ratio A **ratio** is a comparison of the sizes of two quantities of the same kind. It is a single number, usually written as a fraction. For example, the steepness of a hill can be written as the ratio of its height to its horizontal extent.

$$\text{Steepness} = \frac{10 \text{ ft}}{80 \text{ ft}} = \frac{1}{8}$$

The ratio of the circumference of any circle to its diameter is approximately 3.14. This ratio is used so often in mathematics that it has been given a special symbol, the Greek letter π (pronounced "pie").

Proportion A **proportion** is a statement that two ratios are equal. It is an equation or a statement in words and is most interesting when one of the ratios is incomplete. For example, if the ratio of Jill's height to John's height is 7 to 8 and John is 6 feet tall, how tall is Jill?

$$\text{Ratio of heights} = \frac{\text{Jill's height}}{\text{John's height}} = \frac{7}{8} = \frac{\boxed{?}}{6 \text{ ft}}$$

This equation is a proportion. It is always true that if two fractions are equal

$$\frac{a}{b} = \frac{c}{d} \text{ then } \frac{a \times d}{b \times d} = \frac{c \times b}{d \times b}$$

Or, since the denominators are equal, $a \times d = c \times b$. In any proportion the cross products are equal.

$$\frac{a}{b} = \frac{c}{d} \quad a \times d = c \times b$$

Therefore, $\boxed{?} \times 8 = 7 \text{ ft} \times 6 \text{ ft}$

$$\boxed{?} = \frac{7 \text{ ft} \times 6 \text{ ft}}{8 \text{ ft}} = \frac{42}{8} \text{ ft} = 5\frac{1}{4} \text{ ft} = 5 \text{ ft } 3 \text{ in.}$$

The following problem shows how useful proportions can be.

A telephone pole casts a shadow 9 ft long. A yardstick beside it casts a shadow 16 inches long. What is the height of the telephone pole?

$$\frac{[?]}{9 \text{ ft}} = \frac{36 \text{ in.}}{16 \text{ in.}}$$

$$[?] = \frac{36 \text{ in.} \times 9 \text{ ft}}{16 \text{ in.}} = \frac{81}{4} \text{ ft} = 20\frac{1}{4} \text{ ft}$$

Try the following problems to exercise your understanding of ratio and proportion.

1. In the College of Web Technology 2430 students are male and 2970 are female. How many women students would you expect to find in a class where there are 18 men?
2. In the first 12 games of the season, Skye Hooke, our star basketball player, scored 252 points. How many can he expect to score in the entire 22-game season?
3. A Cardiac SUV travels 132 miles on 8 gallons of gasoline. How far would you expect it to travel on a full tank of 12 gallons?

The answers to these problems are in the Appendix.

---

33   *Problem Set 2-3:*   Dividing Fractions

A. Divide and reduce the answer to lowest terms:

1. $\frac{5}{6} \div \frac{1}{2}$ 2. $\frac{3}{4} \div \frac{3}{7}$ 3. $6 \div \frac{2}{3}$

4. $\frac{1}{2} \div 6$ 5. $\frac{5}{12} \div \frac{4}{3}$ 6. $\frac{4}{18} \div \frac{1}{2}$

7. $8 \div \frac{1}{3}$ 8. $\frac{7}{20} \div \frac{4}{5}$ 9. $\frac{6}{13} \div \frac{3}{4}$

10. $3 \div \frac{2}{5}$ 11. $\frac{1}{2} \div \frac{1}{2}$ 12. $\frac{1}{2} \div \frac{1}{3}$

13. $\frac{3}{14} \div \frac{6}{5}$ 14. $\frac{3}{5} \div \frac{1}{3}$ 15. $\frac{3}{4} \div \frac{5}{16}$

B. Divide and reduce the answer to lowest terms:

1. $1\frac{1}{2} \div \frac{1}{6}$ 　　　 2. $2\frac{3}{4} \div \frac{3}{8}$ 　　　 3. $6 \div 1\frac{1}{2}$

4. $2\frac{1}{4} \div 3$ 　　　 5. $3\frac{1}{7} \div 2\frac{5}{14}$ 　　　 6. $8\frac{2}{5} \div 1\frac{2}{5}$

7. $3\frac{1}{2} \div 2$ 　　　 8. $4\frac{1}{2} \div 1\frac{3}{4}$ 　　　 9. $6\frac{2}{5} \div 5\frac{1}{3}$

10. $10 \div 1\frac{1}{5}$ 　　　 11. $4\frac{1}{6} \div 3\frac{1}{3}$ 　　　 12. $15\frac{5}{6} \div 9\frac{1}{2}$

13. $7\frac{1}{7} \div 8\frac{1}{3}$ 　　　 14. $11\frac{2}{3} \div 2\frac{2}{9}$ 　　　 15. $1\frac{1}{5} \div 1\frac{1}{2}$

C. Divide:

1. $\dfrac{8}{\frac{1}{2}}$ 　　 2. $\dfrac{\frac{3}{4}}{2}$ 　　 3. $\dfrac{\frac{2}{3}}{6}$

4. $\dfrac{2\frac{1}{3}}{\frac{3}{4}}$ 　　 5. $\dfrac{12}{\frac{2}{3}}$ 　　 6. $\dfrac{15}{\frac{3}{5}}$

7. Divide $\frac{3}{4}$ by $\frac{7}{8}$. 　　　　 8. Divide 2 by $\frac{1}{3}$.

9. Divide $1\frac{7}{8}$ by $\frac{3}{2}$. 　　　　 10. Divide $1\frac{1}{2}$ by $7\frac{1}{4}$.

11. Divide 5 by $\frac{2}{7}$. 　　　　 12. Divide $11\frac{1}{3}$ by $\frac{2}{3}$.

D. Brain Boosters

1. Mr. Megabuck drove his new stretch SUV 34½ miles and used 3¼ gallons of gasoline. How many miles per gallon did he average?

2. The product of two numbers is 14⅔. One of the numbers is 3⅓. What is the other number?

3. If you drive 59½ miles in 1¾ hours, what is your average speed?

4. **Business**   At the Sew-Rite Fabric Shop, a length of cloth 6⅞ yards long is to be divided into 5 equal pieces. What will be the length of each piece?

5. **Construction**   How many pieces of plywood ½ inch thick are there in a stack 42 inches high?

6. **Business**   How many shares of stock selling for $5⅜ on WebStock.com can be purchased for $2150?

The answers to these problems are in the Appendix. When you have had the practice you need, either return to the preview test on page 83 or continue in **34** with the addition and subtraction of fractions.

4   Adding and Subtracting Fractions

34   At heart, adding fractions is a matter of counting:

$$\frac{1}{5} + \frac{3}{5} = \frac{1+3}{5} = \frac{4}{5}$$

←   1 fifth

+

←   + 3 fifths

=

←   4 fifths (count them)

You try this one:

$$\frac{2}{7} + \frac{3}{7} = \underline{\hspace{2cm}}$$

Check your answer in **35**.

35 $$\frac{2}{7} + \frac{3}{7} = \frac{2+3}{7} = \frac{5}{7}$$

←   2 sevenths

+

←   + 3 sevenths

=

←   5 sevenths or $\frac{5}{7}$

Fractions having the same denominator are called **like fractions**. In the problem at the bottom of page 114, $\frac{2}{7}$ and $\frac{5}{7}$ both have the denominator 7 and are like fractions. Adding like fractions is easy.

To add like fractions:

- **First,** add the numerators to find the numerator of the sum.
- **Second,** use the denominator the fractions have in common as the denominator of the sum.

$$\frac{2}{9} + \frac{5}{9} = \frac{2+5}{9} = \frac{7}{9} \quad \longleftarrow \quad \text{Add numerators}$$
$$\longleftarrow \quad \text{Same denominator}$$

Adding three or more fractions presents no special problems.

$$\frac{3}{12} + \frac{1}{12} + \frac{5}{12} = \underline{\hphantom{xxxx}}$$

Add the fractions as shown above, then turn to **36**.

36 $\quad \dfrac{3}{12} + \dfrac{1}{12} + \dfrac{5}{12} = \dfrac{3+1+5}{12} = \dfrac{9}{12} = \dfrac{3}{4}$

LEARNING HELP ⬦   Always reduce the sum to lowest terms.

Try these problems for practice:

(a) $\dfrac{1}{8} + \dfrac{3}{8} = \underline{\hphantom{xxx}}$ (b) $\dfrac{7}{9} + \dfrac{5}{9} = \underline{\hphantom{xxx}}$

(c) $2\dfrac{1}{2} + 3\dfrac{3}{2} = \underline{\hphantom{xxx}}$ (d) $2 + 3\dfrac{1}{2} = \underline{\hphantom{xxx}}$

(e) $\dfrac{1}{7} + \dfrac{4}{7} + \dfrac{5}{7} + 1\dfrac{2}{7} + \dfrac{8}{7} = \underline{\hphantom{xxx}}$

(f) $\dfrac{3}{5} + 1\dfrac{1}{5} + 3 = \underline{\hphantom{xxx}}$

Go to **38** to check your work.

37 $\quad \dfrac{3}{4} + \dfrac{2}{3} = \dfrac{9}{12} + \dfrac{8}{12} = \dfrac{17}{12} = 1\dfrac{5}{12}$

We change the original fractions to equivalent fractions with the same denominator and then add as before.

How do you know what number to use as the new denominator? In general, you cannot simply guess at the best new denominator. We need a method for finding it from the denominators of the fractions to be added.

If the larger denominator is a multiple of the smaller one, use the larger as the new denominator. For example, to add

$$\frac{1}{8} + \frac{1}{4}$$

first notice that 8 is a multiple of 4. Use 8 as the new denominator.

Rewrite the second fraction as an equivalent fraction with denominator 8.

$$\frac{1}{4} = \frac{1 \times 2}{4 \times 2} = \frac{2}{8}$$

Then,

$$\frac{1}{8} + \frac{1}{4} = \frac{1}{8} + \frac{2}{8} = \frac{3}{8}$$

 Using a calculator,

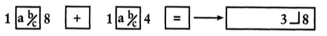

Use this procedure to add the following fractions.

(a) $\dfrac{1}{2} + \dfrac{5}{6}$ (b) $\dfrac{2}{3} + \dfrac{5}{12}$

(c) $\dfrac{3}{16} + \dfrac{1}{4}$ (d) $\dfrac{1}{4} + \dfrac{3}{32}$

Our answers are in **39**.

38 (a) $\dfrac{1}{8} + \dfrac{3}{8} = \dfrac{1+3}{8} = \dfrac{4}{8} = \dfrac{1}{2}$ (reduced to lowest terms)

(b) $\dfrac{7}{9} + \dfrac{5}{9} = \dfrac{7+5}{9} = \dfrac{12}{9} = \dfrac{4}{3} = 1\dfrac{1}{3}$

(c) $2\dfrac{1}{2} + 3\dfrac{3}{2} = 2 + 3 + \dfrac{1}{2} + \dfrac{3}{2} = 2 + 3 + \dfrac{4}{2} = 2 + 3 + 2 = 7$

(Add the whole number parts first.)

(d) $2 + 3\dfrac{1}{2} = 2 + 3 + \dfrac{1}{2} = 5\dfrac{1}{2}$ $\left(\text{Remember, } 3\dfrac{1}{2} \text{ means } 3 + \dfrac{1}{2}.\right)$

(e) $\dfrac{1}{7} + \dfrac{4}{7} + \dfrac{5}{7} + 1\dfrac{2}{7} + \dfrac{8}{7} = \dfrac{1}{7} + \dfrac{4}{7} + \dfrac{5}{7} + \dfrac{9}{7} + \dfrac{8}{7}$

$= \dfrac{1+4+5+9+8}{7} = \dfrac{27}{7} = 3\dfrac{6}{7}$

(f) $\dfrac{3}{5} + 1\dfrac{1}{5} + 3 = 1 + 3 + \dfrac{3}{5} + \dfrac{1}{5} = 4 + \dfrac{4}{5} = 4\dfrac{4}{5}$

(Add the whole number parts first.)

Unlike Fractions How do we add fractions whose denominators are not the same?

$$\dfrac{2}{3} + \dfrac{3}{4} = \underline{\quad ? \quad}$$

The problem is to find a simple numeral that names this new number. One way to find it is to change these fractions to equivalent fractions with the same denominator. (Equivalent fractions were discussed earlier, in frame **10**.)

$$\dfrac{3}{4} = \dfrac{3 \times 3}{4 \times 3} = \dfrac{9}{12} \qquad\qquad \dfrac{2}{3} = \dfrac{2 \times 4}{3 \times 4} = \dfrac{8}{12}$$

Now that ⅔ and ¾ have been changed to equivalent fractions, you can add them. Do it now and then go to **37**.

39 (a) The larger denominator, 6, is a multiple of the smaller, 2. Rewrite the first fraction as an equivalent fraction with denominator 6.

$$\dfrac{1}{2} = \dfrac{1 \times 3}{2 \times 3} = \dfrac{3}{6}$$

Then, $\frac{1}{2} + \frac{1}{6} = \frac{3}{6} + \frac{1}{6} = \frac{4}{6}$, or $\frac{2}{3}$ reduced to lowest terms. The sum is equal to $\frac{2}{3}$ or $1\frac{1}{3}$.

(b) $\dfrac{2}{3} = \dfrac{2 \times 4}{3 \times 4} = \dfrac{8}{12}$

Then the sum is $\frac{2}{3} + \frac{5}{12} = \frac{8}{12} + \frac{5}{12} = \frac{13}{12}$, or $1\frac{1}{12}$.

(c) The denominator 4 is a multiple of 16; therefore,

$$\frac{1}{4} = \frac{1 \times 4}{4 \times 4} = \frac{4}{16}$$

and the sum is $\frac{3}{16} + \frac{4}{16} = \frac{7}{16}$.

(d) $\dfrac{1}{4} = \dfrac{1 \times 8}{4 \times 8} = \dfrac{8}{32}$

and the sum is $\frac{1}{4} + \frac{3}{32} = \frac{8}{32} + \frac{3}{32} = \frac{11}{32}$.

If one denominator is not a multiple of the other, another number must be chosen as the new denominator. This new denominator is called **LCM** the **Least Common Multiple**, or **LCM**, of the given denominators. It is the smallest number that is a multiple of both of the original denominators.

To find the LCM, follow these steps.

Step 1

Write each number as a product of primes.

Example   Find the LCM of 15 and 18.
$15 = 3 \times 5$
$18 = 2 \times 3 \times 3 = 2 \times 3^2$

Step 2

Write each base prime that appears in either number.

2   3   5   ⟵⟶ These are the only primes that appear.

Step 3

Attach to each prime the largest exponent that appears on it in either number.

2^1 3^2 5^1

| 3 appears twice in the factorization of 18. |

Step 4

Multiply to find the LCM.

$$\text{LCM} = 2^1 \times 3^2 \times 5^1$$
$$= 2 \times 9 \times 5$$
$$= 90$$

Practice this by finding the LCM of 12 and 30. Check your work in **40**.

40 The larger number, 30, is not a multiple of the smaller; therefore we must find the LCM of the two numbers.

Step 1 $12 = 2 \times 2 \times 3 = 2^2 \times 3$
            $30 = 2 \times 3 \times 5$

Step 2   The primes are 2, 3, 5.

Step 3   The factor 2 appears twice in 12. The other factors appear only once in each number.

Step 4 $LCM = 2^2 \times 3 \times 5$
                  $= 4 \times 3 \times 5$
                  $= 60$ The LCM of 12 and 30 is 60. This is the smallest number that is a multiple of both 12 and 30.

Find the LCM of each of the following pairs of numbers.

(a) 4 and 6   (b) 8 and 18   (c) 12 and 16

Compare your work with ours in **41**.

41 (a) $4 = 2 \times 2 = 2^2$
             $6 = 2 \times 3$
        $LCM = 2^2 \times 3$ ◄——————— The factor 2 appears twice in 4.
             $= 4 \times 3$
             $= 12$

(b) $8 = 2 \times 2 \times 2 = 2^3$
     $18 = 2 \times 3 \times 3 = 2 \times 3^2$
 $LCM = 2^3 \times 3^2$ ◄——————— The factor 2 appears three times in 8.
     $= 8 \times 9$ The factor 3 appears twice in 18.
     $= 72$

(c) $12 = 2 \times 2 \times 3 = 2^2 \times 3$
     $16 = 2 \times 2 \times 2 \times 2 = 2^4$
 $LCM = 2^4 \times 3$ ◄——————— The factor 2 appears four times in the
     $= 16 \times 3$ number 16. The factor 3 appears only
     $= 48$ once in 12.

Here is a shortcut way of finding the LCM.

Step 1

Write the numbers on a horizontal line.

Example Find the LCM of 12 and 16.

$\lvert 12 \quad 16$

Step 2

Divide both numbers by any factor they have in common. Write the quotients below.

$$2\overline{)\begin{array}{cc} 12 & 16 \\ 6 & 8 \end{array}}$$

Step 3

Repeat the procedure, dividing by any factor that they have in common until the numbers that remain have no common factor.

$$\begin{array}{c|cc} 2 & 12 & 16 \\ 2 & 6 & 8 \\ & 3 & 4 \end{array}$$

Step 4

The LCM is the product of the factors shown.

$$LCM = 2 \times 2 \times 3 \times 4 = 48$$

Try it. Find the LCM of 8 and 20 using this process. Check your work in **42**.

42

$$\begin{array}{c|cc} 2 & 8 & 20 \\ 2 & 4 & 10 \\ & 2 & 5 \end{array}$$

◄——— First, divide by the common factor 2.
◄——— Notice that 2 is also a common factor of 4 and 10.
◄——— 2 and 5 have no common factor.

$$LCM = 2 \times 2 \times 2 \times 5$$
$$= 40$$

Practice using this shortcut by finding the LCM of each of the following pairs of numbers.

(a) 20 and 24   (b) 24 and 36   (c) 60 and 40

Check your work in **43**.

43

(a)
$$\begin{array}{c|cc} 2 & 20 & 24 \\ 2 & 10 & 12 \\ & 5 & 6 \end{array}$$

◄——— Divide by the common factor 2.
◄——— 2 is a factor of both 10 and 12.
◄——— 5 and 6 have no common factors.

$$LCM = 2 \times 2 \times 5 \times 6$$
$$= 120$$

(b)
$$\begin{array}{c|cc} 2 & 24 & 36 \\ 2 & 12 & 18 \\ 3 & 6 & 9 \\ & 2 & 3 \end{array}$$

◄——— Divide by the common factor 2.
◄——— Again 2 is a common factor.
◄——— 3 is a factor of both 6 and 9.
◄——— 2 and 3 have no common factor.

$$LCM = 2 \times 2 \times 3 \times 2 \times 3$$
$$= 72$$

(c) $\begin{array}{c} 10 \end{array}\Big\vert \begin{array}{cc} 60 & 40 \end{array}$ ◄————— 10 is a factor of both 60 and 40.

$2\Big\vert \begin{array}{cc} 6 & 4 \end{array}$ ◄————— Divide by 2, the common factor of 6 and 4.

$\begin{array}{cc} 3 & 2 \end{array}$ ◄————— 3 and 2 have no common factor.

$$LCM = 10 \times 2 \times 3 \times 2$$
$$= 120$$

This same procedure can be used to find the LCM of three or more numbers. For example, find the LCM of 24, 16, and 30 this way.

$2\Big\vert\begin{array}{ccc} 24 & 16 & 30 \end{array}$ ◄————— 2 is a common factor. Divide each by 2.

$\begin{array}{ccc} 12 & 8 & 15 \end{array}$

$2\Big\vert\begin{array}{ccc} 24 & 16 & 30 \end{array}$

$4\Big\vert\begin{array}{ccc} 12 & 8 & 15 \end{array}$ ◄————— 4 is a factor of 12 and 8. Divide them by 4.

$\begin{array}{ccc} 3 & 2 & 15 \end{array}$ ◄————— Since 15 is not divisible by 4, bring it down.

$2\Big\vert\begin{array}{ccc} 24 & 16 & 30 \end{array}$

$4\Big\vert\begin{array}{ccc} 12 & 8 & 15 \end{array}$

$3\Big\vert\begin{array}{ccc} 3 & 2 & 15 \end{array}$ ◄————— 3 is a factor of 3 and 15. Divide them by 3.

$\begin{array}{ccc} 1 & 2 & 5 \end{array}$ — 2 is not divisible by 3; so bring it down.

$$LCM = 2 \times 4 \times 3 \times 1 \times 2 \times 5$$
$$= 240$$

Try it. Find the LCM for these numbers.

(a) 4, 6, 15   (b) 8, 12, 18

Check your work in **44**.

44   (a) $2\Big\vert\begin{array}{ccc} 4 & 6 & 15 \end{array}$ ◄————— 4 and 6 are divisible by 2.

$3\Big\vert\begin{array}{ccc} 2 & 3 & 15 \end{array}$ ◄————— 3 and 15 are divisible by 3.

$\begin{array}{ccc} 2 & 1 & 5 \end{array}$

$$LCM = 2 \times 3 \times 2 \times 1 \times 5$$
$$= 60$$

(b) $2\Big\vert\begin{array}{ccc} 8 & 12 & 18 \end{array}$ ◄————— All are divisible by 2.

$2\Big\vert\begin{array}{ccc} 4 & 6 & 9 \end{array}$ ◄————— 4 and 6 are divisible by 2.

$3\Big\vert\begin{array}{ccc} 2 & 3 & 9 \end{array}$ ◄————— 3 and 9 are divisible by 3.

$\begin{array}{ccc} 2 & 1 & 3 \end{array}$

$$LCM = 2 \times 2 \times 3 \times 2 \times 1 \times 3$$
$$= 72$$

Find the sum of $\frac{3}{18} + \frac{5}{12}$. Write both numbers as equivalent fractions with the LCM of 18 and 12 as the denominator, and then add. Our solution is in **45**.

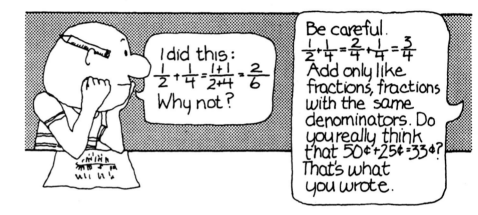

45 $18 = 2^1 \times 3^2$ $12 = 2^2 \times 3$
$LCM = 2^2 \times 3^2 = 4 \times 9 = 36$

$$\frac{3}{18} = \frac{?}{36} = \frac{3 \times 2}{18 \times 2} = \frac{6}{36} \qquad\qquad \frac{5}{12} = \frac{?}{36} = \frac{5 \times 3}{12 \times 3} = \frac{15}{36}$$

Add: $\dfrac{3}{18} + \dfrac{5}{12} = \dfrac{6}{36} + \dfrac{15}{36} = \dfrac{6+15}{36} = \dfrac{21}{36} = \dfrac{7}{12}$ (in lowest terms)

 And of course it is much quicker using a calculator:

$3 \boxed{a\,b/c}\ 18\ \boxed{+}\ 5\boxed{a\,b/c}\ 12\ \boxed{=}\ \longrightarrow \boxed{\qquad 7 \rfloor 12}$

LEARNING HELP ⬦   To add fractions, remember these steps:

1. Add the whole number parts of the fractions.

2. Find the LCM of the denominators of the fractions to be added.

3. Write the fractions so that they are equivalent fractions with the LCM as denominator.

4. The numerator of the sum is the sum of the numerators, and the denominator of the sum is the LCM.

5. Reduce to lowest terms.

This way of adding fractions may seem rather long and involved. It *is* involved, but it is the only sure way to arrive at the answer without using a calculator. Of course, if one is available, a calculator with a fraction key will make these calculations easy, but it is very important that you understand the basic process and that you check your answers.

Practice by adding the following:

(a) $\dfrac{5}{18} + \dfrac{5}{16}$
  (b) $\dfrac{7}{12} + \dfrac{3}{16}$

(c) $\dfrac{10}{32} + \dfrac{4}{30}$
  (d) $\dfrac{3}{7} + \dfrac{4}{5}$

(e) $1\dfrac{2}{15} + \dfrac{5}{9}$
  (f) $\dfrac{1}{2} + \dfrac{5}{6} + \dfrac{3}{4}$

(g) $2\dfrac{5}{12} + 1\dfrac{1}{9} + 2\dfrac{3}{8}$
  (h) $\dfrac{13}{16} + 2\dfrac{1}{18} + 1\dfrac{11}{12}$

The worked solutions are in **48**.

46
$$\frac{3}{8} - \frac{1}{8} = \frac{3-1}{8} = \frac{2}{8} = \frac{1}{4}$$

Easy enough? If the denominators are the same, we subtract numerators and write this difference over the common denominator. Try another:

$$\frac{3}{4} - \frac{1}{5} = \underline{\qquad}$$

Our solution is in **47**.

47 The LCM of 4 and 5 is 20.

$$\frac{3}{4} = \frac{?}{20} = \frac{3 \times 5}{4 \times 5} = \frac{15}{20} \qquad\qquad \frac{1}{5} = \frac{?}{20} = \frac{1 \times 4}{5 \times 4} = \frac{4}{20}$$

Then

$$\frac{3}{4} - \frac{1}{5} = \frac{15}{20} - \frac{4}{20} = \frac{11}{20}$$

When the two fractions have different denominators, we must change them to equivalent fractions with the LCM as denominator before subtracting.

If you have not yet learned how to find the Least Common Multiple (LCM), turn to **39**. Otherwise, continue by finding the following differences.

(a) $3\frac{1}{4} - 1\frac{1}{12} = $ _____

(b) $10 - 3\frac{5}{16} = $ _____

(c) $4\frac{7}{18} - 2\frac{11}{12} = $ _____

(d) $2\frac{6}{32} - 1\frac{1}{6} = $ _____

Check your answers in **49**.

What Is the LCM?

An LCM is not a *Little Crooked Martian*, but a *Least Common Multiple*. A *multiple* of a whole number is a number evenly divisible by it.

4, 8, 12, 16, 20, 24, . . . are all multiples of 4

A pair of numbers will have *common multiples*.

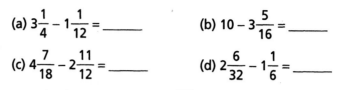

The multiples of 4 are 4, 8, 12, 16, 20, 24, ...
The multiples of 8 are 8, 16, 24, 32, ...
The common multiples of 4 and 8 are 8, 16, 24, and so on.

The smallest common multiple, or *least common multiple* or LCM, is the smallest of all the common multiples of a pair of whole numbers.

The LCM of 4 and 8 is 8. The LCM 8 is the smallest whole number that is evenly divisible by both 4 and 8.

---

48 (a) The LCM of 18 and 16 is 144.

$$\frac{5}{18} = \frac{?}{144} = \frac{5 \times 8}{18 \times 8} = \frac{40}{144} \qquad \frac{5}{16} = \frac{?}{144} = \frac{5 \times 9}{16 \times 9} = \frac{45}{144}$$

$$\frac{5}{18} + \frac{5}{16} = \frac{40}{144} + \frac{45}{144} = \frac{40 + 45}{144} = \frac{85}{144}$$

(b) The LCM of 12 and 16 is 48.

$$\frac{7}{12} = \frac{7 \times 4}{12 \times 4} = \frac{28}{48} \qquad \frac{3}{16} = \frac{3 \times 3}{16 \times 3} = \frac{9}{48}$$

$$\frac{7}{12} + \frac{3}{16} = \frac{28}{48} + \frac{9}{48} = \frac{37}{48}$$

(c) The LCM of 32 and 30 is 480.

$$\frac{10}{32} = \frac{10 \times 15}{32 \times 15} = \frac{150}{480} \qquad \frac{4}{30} = \frac{4 \times 16}{30 \times 16} = \frac{64}{480}$$

$$\frac{10}{32} + \frac{4}{30} = \frac{150}{480} + \frac{64}{480} = \frac{214}{480} = \frac{107}{240}$$

(d) The LCM of 7 and 5 is 35.

$$\frac{3}{7} = \frac{3 \times 5}{7 \times 5} = \frac{15}{35} \qquad \frac{4}{5} = \frac{4 \times 7}{5 \times 7} = \frac{28}{35}$$

$$\frac{3}{7} + \frac{4}{5} = \frac{15}{35} + \frac{28}{35} = \frac{43}{35} = 1\frac{8}{35}$$

(e) The LCM of 15 and 9 is 45.

$$\frac{2}{15} = \frac{2 \times 3}{15 \times 3} = \frac{6}{45} \qquad \frac{5}{9} = \frac{5 \times 5}{9 \times 5} = \frac{25}{45}$$

$$1\frac{2}{15} + \frac{5}{9} = 1 + \frac{2}{15} + \frac{5}{9} = 1 + \frac{6}{45} + \frac{25}{45} = 1 + \frac{31}{45} = 1\frac{31}{45}$$

(f) The LCM of 2, 6, and 4 is 12.

$$\frac{1}{2} = \frac{6}{12} \qquad \frac{5}{6} = \frac{10}{12} \qquad \frac{3}{4} = \frac{9}{12}$$

$$\frac{1}{2} + \frac{5}{6} + \frac{3}{4} = \frac{6}{12} + \frac{10}{12} + \frac{9}{12} = \frac{25}{12} = 2\frac{1}{12}$$

(g) The LCM of 12, 9, and 8 is 72.

$$\frac{5}{12} = \frac{5 \times 6}{12 \times 6} = \frac{30}{72} \qquad \frac{1}{9} = \frac{1 \times 8}{9 \times 8} = \frac{8}{72} \qquad \frac{3}{8} = \frac{3 \times 9}{8 \times 9} = \frac{27}{72}$$

$$2\frac{5}{12} + 1\frac{1}{9} + 2\frac{3}{8} = 2 + 1 + 2 + \frac{5}{12} + \frac{1}{9} + \frac{3}{8} = 5 + \frac{30}{72} + \frac{8}{72} + \frac{27}{72}$$

$$= 5 + \frac{65}{72} = 5\frac{65}{72}$$

(h) The LCM of 16, 18, and 12 is 144.

$$\frac{13}{16} = \frac{117}{144} \qquad \frac{1}{18} = \frac{8}{144} \qquad \frac{11}{12} = \frac{132}{144}$$

$$\frac{13}{16} + 2\frac{1}{18} + 1\frac{11}{12} = 2 + 1 + \frac{117}{144} + \frac{8}{144} + \frac{132}{144} = 3 + \frac{257}{144}$$

$$= 4 + \frac{113}{144} = 4\frac{113}{144}$$

Once you have mastered the process of adding fractions, subtraction is very simple indeed. Try this problem:

$$\frac{3}{8} - \frac{1}{8} = \underline{\qquad}$$

Turn to .

49   (a) The LCM of 4 and 12 is 12.

$$3\frac{1}{4} = \frac{13}{4} = \frac{13 \times \boxed{3}}{4 \times \boxed{3}} = \frac{39}{12} \qquad\qquad 1\frac{1}{12} = \frac{13}{12}$$

$$3\frac{1}{4} - 1\frac{1}{12} = \frac{39}{12} - \frac{13}{12} = \frac{26}{12} = \frac{13}{6} = 2\frac{1}{6}$$

(b) $10 = \dfrac{10}{1} = \dfrac{10 \times \boxed{16}}{1 \times \boxed{16}} = \dfrac{160}{16} \qquad\qquad 3\dfrac{5}{16} = \dfrac{53}{16}$

$$10 - 3\frac{5}{16} = \frac{160}{16} - \frac{53}{16} = \frac{107}{16} = 6\frac{11}{16}$$

(c) The LCM of 18 and 12 is 36.

$$4\frac{7}{18} = \frac{79}{18} = \frac{79 \times \boxed{2}}{18 \times \boxed{2}} = \frac{158}{36} \qquad\qquad 2\frac{11}{12} = \frac{35}{12} = \frac{35 \times \boxed{3}}{12 \times \boxed{3}} = \frac{105}{36}$$

$$4\frac{7}{18} - 2\frac{11}{12} = \frac{158}{36} - \frac{105}{36} = \frac{53}{36} = 1\frac{17}{36}$$

(d) The LCM of 32 and 6 is 96.

$$2\frac{6}{32} = \frac{70}{32} = \frac{70 \times \boxed{3}}{32 \times \boxed{3}} = \frac{210}{96} \qquad\qquad 1\frac{1}{6} = \frac{7}{6} = \frac{7 \times \boxed{16}}{6 \times \boxed{16}} = \frac{112}{96}$$

$$2\frac{6}{32} - 1\frac{1}{6} = \frac{210}{96} - \frac{112}{96} = \frac{98}{96} = \frac{49}{48} = 1\frac{1}{48}$$

Now turn to **50** for a set of practice problems on addition and subtraction of fractions.

50   *Problem Set 2-4:*   Adding and Subtracting Fractions

A. Add or subtract as shown:

1. $\dfrac{3}{5} + \dfrac{4}{5}$ 2. $\dfrac{5}{12} + \dfrac{11}{12}$ 3. $\dfrac{1}{8} + \dfrac{7}{8}$

4. $\dfrac{7}{15} + \dfrac{3}{15}$ 5. $\dfrac{7}{8} - \dfrac{5}{8}$ 6. $\dfrac{3}{4} - \dfrac{1}{4}$

7. $\dfrac{7}{12} - \dfrac{2}{12}$ 8. $\dfrac{15}{16} - \dfrac{9}{16}$ 9. $\dfrac{53}{64} + \dfrac{19}{64}$

10. $\dfrac{17}{32} + \dfrac{19}{32}$ 11. $\dfrac{53}{64} - \dfrac{5}{64}$ 12. $\dfrac{15}{32} - \dfrac{7}{32}$

13. $\dfrac{2}{16} + \dfrac{7}{16} + \dfrac{13}{16}$ 14. $\dfrac{7}{18} + \dfrac{10}{18} + \dfrac{13}{18}$

15. $\dfrac{1}{7} + \dfrac{6}{7} + \dfrac{4}{7}$ 16. $\dfrac{7}{8} + \dfrac{1}{8} - \dfrac{3}{8}$

17. $\dfrac{5}{12} + \dfrac{11}{12} - \dfrac{1}{12}$ 18. $\dfrac{7}{20} - \dfrac{11}{20} + \dfrac{13}{20}$

B. Add or subtract as shown:

1. $\dfrac{7}{8} + \dfrac{3}{4}$ 2. $\dfrac{7}{8} - \dfrac{3}{4}$ 3. $\dfrac{11}{12} + \dfrac{7}{18}$

4. $\dfrac{11}{12} - \dfrac{7}{18}$ 5. $\dfrac{5}{12} + \dfrac{3}{16}$ 6. $\dfrac{3}{7} + \dfrac{2}{5}$

7. $\dfrac{11}{48} + \dfrac{3}{64}$ 8. $\dfrac{10}{27} + \dfrac{15}{24}$ 9. $\dfrac{7}{12} - \dfrac{5}{16}$

10. $\dfrac{5}{7} - \dfrac{3}{5}$ 11. $\dfrac{18}{64} - \dfrac{11}{48}$ 12. $\dfrac{16}{27} - \dfrac{5}{24}$

13. $1\dfrac{1}{4} + \dfrac{3}{8}$ 14. $2\dfrac{3}{4} + 1\dfrac{5}{18}$ 15. $3\dfrac{5}{12} + 1\dfrac{15}{16}$

16. $3\dfrac{5}{8} + 1\dfrac{2}{7}$ 17. $3\dfrac{5}{8} - 2\dfrac{7}{8}$ 18. $3\dfrac{5}{12} - 1\dfrac{15}{16}$

C. Calculate:

1. $2 - \dfrac{1}{3}$

2. $6 - 4\dfrac{3}{16}$

3. $8 - \dfrac{11}{4}$

4. $3 - 1\dfrac{1}{5}$

5. $6\dfrac{1}{2} + 5\dfrac{3}{4} + 8\dfrac{1}{2}$

6. $\dfrac{7}{8} - 1\dfrac{1}{3} + 2\dfrac{1}{5}$

7. $\dfrac{1}{2} + \dfrac{1}{5} + \dfrac{1}{8}$

8. $\dfrac{1}{2} + \dfrac{1}{3} + \dfrac{1}{4} + \dfrac{1}{5}$

9. $1\dfrac{2}{3}$ subtracted from $4\dfrac{3}{4}$

10. $2\dfrac{3}{8}$ less than $4\dfrac{7}{10}$

11. $6\dfrac{5}{12}$ reduced by $1\dfrac{3}{16}$

12. $4\dfrac{3}{5}$ less than $6\dfrac{1}{2}$

D. **Brain Boosters**

1. **Machinist**   Tony, the machinist's helper, got confused and wrote some numbers incorrectly. How much longer is a 1⅝-inch bolt than a 1⅞-inch bolt?

2. **Manufacturing**   Sara's time card at the Sew 'n Sew pajama factory shows that she worked the following hours last week: Monday, 7¼ hours; Tuesday, 6½ hours; Wednesday, 5¾ hours; Thursday, 8 hours; Friday, 9⅙ hours. What total time did she work?

3. Show that:

$$\left(6 + \dfrac{1}{4}\right) \times \left(5 - \dfrac{1}{5}\right) = 6 \times 5$$

$$\left(7 + \dfrac{3}{7}\right) \times \left(4 - \dfrac{3}{13}\right) = 7 \times 4$$

$$\left(31 + \dfrac{1}{2}\right) \times \left(21 - \dfrac{1}{3}\right) = 31 \times 21$$

4. The ancient Egyptians, thousands of years ago, wrote all fractions as a sum of different fractions with numerator 1. For example,

$\dfrac{3}{4}$ would be written $\dfrac{1}{2} + \dfrac{1}{4}$

$\frac{2}{3}$ would be written $\frac{1}{2} + \frac{1}{6}$ $\left(\text{not } \frac{1}{3} + \frac{1}{3}\right)$

How would the Egyptians have written these fractions?

(a) $\frac{7}{8}$ (b) $\frac{5}{9}$ (c) $\frac{5}{12}$

5. **Real Estate** Radis Realty is selling a plot of land whose four sides are $120\frac{3}{4}$ ft, $85\frac{5}{8}$ ft, $116\frac{2}{3}$ ft, and $91\frac{5}{6}$ ft. What is the total distance around the edge of this lot?

6. On a recent trip in his Eletro-Hybrid car, Jose traveled 1846 miles and stopped for gas three times, using $8\frac{1}{2}$ gallons, $10\frac{3}{5}$ gallons, and $9\frac{3}{10}$ gallons.

   (a) How much gas was used?
   (b) What was his average miles per gallon?

7. **Business** Stock in the on-line grocery store Celery.com opened at $47\frac{3}{8}$ yesterday on the New York Stalk Exchange and closed at $45\frac{3}{16}$. What was its net loss in price?

8. Which is larger?

   (a) One-quarter of the sum of $\frac{1}{4}$ plus $\frac{1}{4}$ of $\frac{1}{4}$
   (b) $\frac{1}{4}$ minus $\frac{1}{4}$ of $\frac{1}{4}$

9. **Business** The Zippy Delivery driver drove $8\frac{3}{4}$ hr on Tuesday, $5\frac{1}{2}$ hr on Wednesday, and $9\frac{1}{4}$ hr on Thursday. How many hours did the Zippy driver drive during those three days?

10. **Real Estate** Lots-of-Plots, the home building company, originally owned $234\frac{1}{4}$ acres of land. After selling $145\frac{7}{8}$ acres, how many acres of land did it own?

The answers to these problems are in the Appendix. When you have had the practice you need, either return to the review test on page 83 or continue in 51 with the study of word problems involving fractions.

5 Solving Word Problems

51 The mathematics you use every day in your work or at play usually appears wrapped in words and hidden in sentences. Neat little sets of directions are seldom attached; no "Divide and reduce to lowest terms" or "Write as equivalent fractions and add." The difficulty with real problems

is that they must be translated from words to mathematics. You need to learn to *talk* arithmetic, not just juggle numbers.

Certain words and phrases appear again and again in arithmetic. They are signals alerting you to the mathematical operations to be done. Here is a list of such **signal words.**

Signal Words	**Translate as**
Is, is equal to, equals, the same as	=
Of, the product of, multiply, times, multiplied by	×
Add, in addition, plus, more, more than, sum, and, increased by, added to	+
Subtract, subtract from, less, less than, difference, diminished by, decreased by	−
Divide, divided by	÷
Twice, double, twice as much	2 ×
Half of, half	½ ×

Let's translate the phrase "six times some number." First, make a word equation by using parentheses:

(six) (times) (some number)

Second, translate signal words to math symbols and use □ or a letter of the alphabet to represent any unknown quantity. (The little box □ is handy because you can write in it if you ever learn its numerical value.)

(6) (×) (□)

Translate the phrase "a number divided by seven." Check your answer in **52**.

52   "A number divided by seven" → (a number) (divided by) (seven)

$$\rightarrow \square \div 7 \quad \text{or} \quad \frac{\square}{7} \quad 7\overline{)\square}$$

Translate the following phrases to mathematical expressions.

(a) "Three more than some number"

(b) "One-half of some quantity"

(c) "The sum of ⅔ and a number"

(d) "Six less than a number"

(e) "Five more than twice a number"

(f) "A number divided by ½"

(g) "One-third of a quantity is equal to ¾"

(h) "A number diminished by two-thirds"

Our translations are in **53**.

53 (a) $3 + \square$ (b) $\frac{1}{2} \times \square$ (c) $\frac{2}{3} + \square$

(d) $\square - 6$ (e) $5 + (2 \times \square)$ (f) $\square \div \frac{1}{2}$

(g) $\frac{1}{3} \times \square = \frac{3}{4}$ (h) $\square - \frac{2}{3}$

Several kinds of word problems involving fractions appear again and again. Translate these:

(a) What fraction of 3¾ is 1⅕?

(b) ⅗ of what number is 1⅔?

(c) ⅘ of 1⅞ is what number?

Translate as before, and then turn to **54** for our answers.

54 (a) (What fraction) (of) $\left(3\frac{3}{4}\right)$ (is) $\left(1\frac{1}{5}\right)$?

$$\downarrow \qquad \downarrow \quad \downarrow \quad \downarrow \quad \downarrow$$

$$\square \quad \times \quad 3\frac{3}{4} \quad = \quad 1\frac{1}{5} \qquad \text{or} \qquad \square \times 3\frac{3}{4} = 1\frac{1}{5}$$

(b) $\left(\frac{3}{5}\right)$ (of) (what number) (is) $\left(1\frac{2}{3}\right)$?

$$\downarrow \quad \downarrow \qquad \downarrow \qquad \downarrow \quad \downarrow$$

$$\frac{3}{5} \quad \times \qquad \square \qquad = \quad 1\frac{2}{3} \qquad \text{or} \qquad \frac{3}{5} \times \square = 1\frac{2}{3}$$

(c) $\left(\frac{4}{9}\right)$ (of) $\left(1\frac{7}{8}\right)$ (is) (what number)?

$$\downarrow \quad \downarrow \quad \downarrow \quad \downarrow \qquad \downarrow$$

$$\frac{4}{9} \quad \times \quad 1\frac{7}{8} \quad = \qquad \square \qquad \text{or} \qquad \frac{4}{9} \times 1\frac{7}{8} = \square$$

Problem (c) is easy to solve. If you do not remember how to do it, you should return to **21** and refresh your memory. Let's concentrate on the other two.

Can you solve problem (a)? It will help if you make up a similar problem, one that you can solve easily, and compare them. For example, "What number multiplied by 3 equals 6?"

$$\square \times 3 = 6$$

Of course the answer is 2. How did you get that answer? It didn't simply pop into your head. Your brain did a quick calculation that went something like this:

"What times 3 gives 6?"
   "Hmm. I know that 6 divided by 3 is 2. Is 2 the answer?"
"Try it: does 2 times 3 give 6?"
   "Yep. The answer must be 2."

In math language your brain did this:

$\Box \times 3 = 6$

$\Box = 6 \div 3 = 2$

Check: $\boxed{2} \times 3 = 6$. OK.

Now solve $\Box \times 3\frac{3}{4} = 1\frac{1}{5}$ in exactly the same way you solved $\Box \times 3 = 6$. Check your work in **55**.

55

$\Box \times 3\frac{3}{4} = 1\frac{1}{5}$ **Similar Problem**

$\Box = 1\frac{1}{5} \div 3\frac{3}{4}$ $\Box \times 3 = 6$

$\Box = \frac{6}{5} \div \frac{15}{4}$ $\Box = 6 \div 3$

$\Box = \frac{6}{5} \times \frac{4}{15}$ $\Box = 2$

$\Box = \frac{8}{25}$ Check: $\boxed{2} \times 3 = 6$

Check: $\boxed{\frac{8}{25}} \times 3\frac{3}{4} = 1\frac{1}{5}$ $6 = 6$

$\frac{8}{25} \times \frac{15}{4} = \frac{6}{5} = 1\frac{1}{5}$

LEARNING HELP ⭢ The very best way to solve problems of this kind is to:

1. **Make up** a **similar** problem using small whole numbers—an easy problem that you can solve immediately.

2. **Solve** the difficult problem in exactly the same way.

3. **Check** your answer by substituting it for the unknown quantity in the original equation.

The similar problem you make up allows you to use your feel for the logic of the problem to help you decide how to do it. You get to "think it out" without letting numbers get in the way. Don't hesitate on this; learn to trust yourself—but check every answer.

Now try to solve $\square \times 1\frac{7}{8} = \frac{5}{16}$. The solution is in **57**.

56 $\left(\frac{3}{4}\right)$ (of) (what number) (is) $\left(1\frac{2}{3}\right)$

$\downarrow \quad \downarrow \qquad \downarrow \qquad \downarrow \quad \downarrow$

$$\frac{3}{4} \quad \times \quad \square \quad = \quad 1\frac{2}{3} \qquad \text{or } \frac{3}{4} \times \square = 1\frac{2}{3}$$

Make up a similar, but very easy, problem: **Similar Problem**
"Two times what number equals six?" $2 \times \square = 6$

$= 6 \div 2$

The answer is 3. How did you get it? $= 3$

$\frac{3}{4} \times \square = 1\frac{2}{3}$ Check: $2 \times \boxed{3} = 6$

$\square = 1\frac{2}{3} \div \frac{3}{4}$ $6 = 6$

$\square = \frac{5}{3} \times \frac{4}{3}$

$\square = \frac{20}{9}$

Check: $\frac{3}{4} \times \boxed{\frac{20}{9}} = \frac{5}{3} = 1\frac{2}{3}$

LEARNING HELP ▷ If you are one of those people who like to memorize things, it may help you to remember that:

If $A \times B = C$

then $B = C \div A$

and $A = C \div B$

(for any numbers A, B, and C not equal to zero)

As a general rule, it is better not to try to memorize a rule for every variety of math problem and far better to develop your ability to use basic principles, but if you like to memorize helpful rules this is a very useful one.

Use these problems for practice. Translate each into an equation and solve it.

(a) ⅜ of what number is equal to 1⁵⁄₁₆?

(b) What fraction of 4½ is 6¾?

(c) Find a number such that ³⁄₁₁ of it is 2⅖.

(d) The product of 1⅞ and 2⅓ is what number?

(e) What fraction of 8¾ is ⁷⁄₁₂?

The answers are in **58**.

 This is similar to the problem "What number times 4 equals 8?"

Similar Problem

$$\square \times 1\frac{7}{8} = \frac{5}{16}$$
$$\square \times 4 = 8$$

$$\square = \frac{5}{16} \div 1\frac{7}{8}$$
$$= 8 \div 4$$

$$\square = \frac{5}{16} \div \frac{15}{8}$$
$$= 2$$

$$\square = \frac{5}{16} \times \frac{8}{15}$$

Check: $\boxed{2} \times 4 = 8$ OK.

$$\square = \frac{1}{6}$$

Check: $\boxed{\frac{1}{6}} \times 1\frac{7}{8} = \frac{5}{16}$

$$\frac{1}{6} \times \frac{15}{8} = \frac{5}{16} \qquad \text{OK.}$$

Isn't it easier when you have the simple problem as a guideline?

Try another problem:

¾ of what number is 1⅔?

Translate this word phrase into an equation in symbols and solve. Check your work in **56**.

58 (a) $\left(\dfrac{3}{8}\right)$ (of) (what number) (is equal to) $\left(1\dfrac{5}{16}\right)$?

$$\frac{3}{8} \times \square = 1\frac{5}{16}$$

$$\square = 1\frac{5}{16} \div \frac{3}{8}$$

$$\square = \frac{21}{16} \times \frac{8}{3}$$

$$\square = \frac{7}{2} = 3\frac{1}{2} \qquad \text{Check:} \quad \frac{3}{8} \times \boxed{\frac{7}{2}} = \frac{21}{16} = 1\frac{5}{16}$$

(b) (What fraction) (of) $\left(4\dfrac{1}{2}\right)$ (is) $\left(6\dfrac{3}{4}\right)$?

$$\square \times 4\frac{1}{2} = 6\frac{3}{4}$$

$$\square = 6\frac{3}{4} \div 4\frac{1}{2}$$

$$\square = \frac{27}{4} \times \frac{2}{9}$$

$$\square = \frac{3}{2} = 1\frac{1}{2} \qquad \text{Check:} \quad \boxed{\frac{3}{2}} \times 4\frac{1}{2} = \frac{3}{2} \times \frac{9}{2} = \frac{27}{4} = 6\frac{3}{4}$$

(c) Find (a number) such that $\left(\dfrac{3}{11}\right)$ (of) (it) (is) $\left(2\dfrac{2}{5}\right)$.

$$\frac{3}{11} \times \square = 2\frac{2}{5}$$

$$\square = 2\frac{2}{5} \div \frac{3}{11}$$

$$\square = \frac{12}{5} \times \frac{11}{3}$$

$$\square = \frac{44}{5} = 8\frac{4}{5} \qquad \text{Check:} \quad \frac{3}{11} \times \boxed{\frac{44}{5}} = \frac{12}{5} = 2\frac{2}{5}$$

(d) The (product of) $\left(1\frac{7}{8}\right)$ and $\left(2\frac{1}{3}\right)$ (is) (what number)?

$$1\frac{7}{8} \times 2\frac{1}{3} = \square$$

$$\frac{15}{8} \times \frac{7}{3} = \square$$

$$\frac{35}{8} = \square$$

$$\square = 4\frac{3}{8}$$

(e) (What fraction) (of) $\left(8\frac{3}{4}\right)$ (is) $\left(\frac{7}{12}\right)$?

$$\square \times 8\frac{3}{4} = \frac{7}{12}$$

$$\square = \frac{7}{12} \div 8\frac{3}{4}$$

$$\square = \frac{7}{12} \div \frac{35}{4}$$

$$\square = \frac{7}{12} \times \frac{4}{35}$$

$$\square = \frac{1}{15} \qquad \text{Check:} \quad \boxed{\frac{1}{15}} \times 8\frac{3}{4} = \frac{1}{15} \times \frac{35}{4} = \frac{7}{12}$$

Here is another type of problem involving fractions:

If $6\frac{2}{3}$ dozen doorknobs cost \$100, what will 10 dozen doorknobs cost?

To solve this, first make up a similar and much easier problem in order to get the feel of it. For example,

If 2 apples cost 10¢, what will 6 apples cost?
Solution: 2 cost 10¢
    1 costs 10¢ ÷ 2 = 5¢
    then 6 cost 6 × 5¢ = 30¢

In other words we solve the problem by dividing it into two parts:

• **First,** find the unit cost, the cost of one item.
• **Second,** find the cost of any number of items.

Apply this method to the doorknob problem. Work exactly as you did in the simpler problem. Check your work in **59**.

59 $6\frac{2}{3}$ cost $100

$$1 \text{ costs} \left(\$100 \div 6\frac{2}{3}\right) = \$100 \div \frac{20}{3}$$

$$= \$100 \times \frac{3}{20} = \$15$$

Then 10 cost $(10 \times \$15) = \$150$

One way to **check** your answer is to compare it with what you might have guessed the answer to be before you did the arithmetic. For example, if 6⅔ cost $100, then 10 will cost almost double that and the answer will be between $100 and $200. The actual answer $150 is reasonable.

Here are a few problems for you to practice with.

(a) If 4½ dozen pencils cost 2¼ dollars, what will 8½ dozen cost?

(b) If you walk 24 miles in 7½ hours, how many miles will you walk in 10 hours, assuming you go at the same rate?

(c) If you can swim ⅞ of one mile in ⅔ hour, how far can you swim in 2½ hours at the same rate?

Look in **60** for the answers.

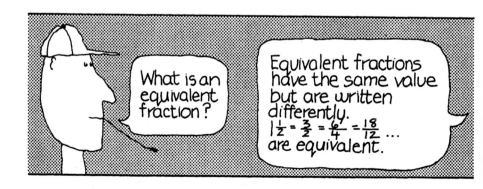

60 (a) $4\frac{1}{2}$ cost $\$2\frac{1}{4}$ **Similar Problem**

1 costs $\$2\frac{1}{4} \div 4\frac{1}{2} = \frac{\$9}{4} \div \frac{9}{2}$ 2 cost 10¢

$= \frac{\$9}{4} \times \frac{2}{9} = \frac{\$1}{2}$ 1 costs 10¢ ÷ 2 = 5¢

Then $8\frac{1}{2}$ cost $8\frac{1}{2} \times \frac{\$1}{2} = \frac{17}{2} \times \frac{\$1}{2}$ 6 cost 6 × 5¢ = 30¢

$= \frac{17}{4} = \$4\frac{1}{4}$

$8\frac{1}{2}$ dozen cost $\$4\frac{1}{4}$

(b) $7\frac{1}{2}$ hours → 24 miles

1 hour → $24 \div 7\frac{1}{2}$

$= 24 \times \frac{2}{15} = \frac{16}{5} = 3\frac{1}{5}$ miles

Then 10 hours → $10 \times \frac{16}{5} = 32$ miles

(c) $\frac{2}{3}$ hour for $\frac{7}{8}$ mile

1 hour → $\frac{7}{8} \div \frac{2}{3} = \frac{7}{8} \times \frac{3}{2} = \frac{21}{16}$ mile

$2\frac{1}{2}$ hours → $\frac{21}{16} \times \frac{5}{2} = \frac{105}{32} = 3\frac{9}{32}$ miles

Now go to **61** for some practice on word problems involving fractions.

Problem Set 2-5: Solving Word Problems

A. Translate these words and phrases to mathematical expressions and symbols:

1. is

2. of

3. increased by

4. the same as

5. 6 subtracted from some number

6. half of a number

7. double the number

8. a number divided by $2\frac{1}{2}$

9. a number plus $\frac{2}{3}$

10. the sum of a number and $\frac{2}{5}$

11. a number divided by $\frac{3}{7}$

12. $\frac{7}{8}$ of a number is equal to $1\frac{1}{2}$

13. some number less $1\frac{3}{4}$

14. What fraction of $3\frac{1}{4}$ is $11\frac{1}{2}$?

15. the product of $\frac{7}{16}$ and $3\frac{5}{8}$

B. Solve:

1. What fraction of $3\frac{1}{4}$ is $4\frac{7}{8}$?

2. What part of $7\frac{1}{2}$ is $2\frac{1}{4}$?

3. What fraction of $3\frac{1}{8}$ is 5?

4. $\frac{7}{8}$ of what number is $\frac{2}{3}$?

5. $\frac{4}{11}$ of what number is $\frac{1}{2}$?

6. $3\frac{1}{8}$ of $\frac{2}{5}$ is what number?

7. Find a number such that $\frac{5}{16}$ of it is $4\frac{1}{2}$.

C. Solve:

1. **Manufacturing** If 4⅞ pounds of Geedunk gear packing cost $39, what will be the cost of 6½ pounds?

2. Des, the Hash House flash, runs 6¼ miles in 35 minutes. How many miles will he cover in 45½ minutes at this pace?

3. On a vacation trip, 25⅗ gallons of gas were used to drive 480 miles. How many gallons were used to drive the first 100 miles at that rate?

4. **Carpentry** If a box containing 2¾ pounds of nails costs $8, how many pounds can be purchased for $12?

5. **Business** If you are paid $138 for 34½ hours of work, what should you be paid for 46½ hours of work at the same rate of pay?

The answers to these problems are in the Appendix. When you have had the practice you need, turn to **62** for a self-test on fractions.

62 CHAPTER 2 SELF-TEST

1. Write $7\dfrac{3}{16}$ as an improper fraction: _____

2. Write $\dfrac{37}{11}$ as a mixed number: _____

3. $\dfrac{3}{8} = \dfrac{?}{40}$ _____

4. Reduce to lowest terms: $\dfrac{195}{255}$

5. $\dfrac{3}{5} + \dfrac{2}{7} =$ _____

6. $\dfrac{1}{4} + \dfrac{2}{3} + \dfrac{2}{5} =$ _____

7. $1\dfrac{3}{8} + 2\dfrac{1}{4} + 2\dfrac{2}{3} =$ _____

8. $\dfrac{3}{4} - \dfrac{1}{3} =$ _____

9. $2\dfrac{2}{5} - 1\dfrac{1}{4} =$ _____

10. $6\dfrac{2}{3} - 3\dfrac{1}{4} =$ _____

11. $\dfrac{9}{15} \times \dfrac{5}{3} =$ _____

12. $2\dfrac{2}{7} \times 2\dfrac{1}{4} =$ _____

13. $\dfrac{2}{3} \div \dfrac{3}{5} =$ _____

14. $3\dfrac{2}{7} \div 7\dfrac{1}{3} =$ _____

15. $1\dfrac{3}{16} \div 4\dfrac{3}{4} =$ _____

16. $\left(2\dfrac{1}{3}\right)^2 =$ _____

17. $1\dfrac{3}{5} \times 4\dfrac{7}{8} \times 7\dfrac{1}{2} =$ _____

18. What fraction of 16 is 7? _____

19. What fraction of $7\dfrac{1}{2}$ is 3? _____

20. What fraction of $4\dfrac{2}{3}$ is $3\dfrac{1}{2}$? _____

21. $\dfrac{7}{8}$ of what number is $1\dfrac{3}{4}$? _____

22. $1\dfrac{2}{3}$ of what number is $\dfrac{7}{15}$? _____

23. Find a number such that $\dfrac{2}{7}$ of it is $3\dfrac{1}{2}$. _____

24. What fraction of 50 is $4\dfrac{1}{2}$? _____

25. **Construction**   If $1\dfrac{2}{3}$ cubic feet of gravel cost $60, what will 2 cubic feet cost? _____

The answers to these problems are in the Appendix.

3 Decimals

Where to Go for Help
Page   **Frame**

3. Convert fractions to decimals and decimals
to fractions.

(a) Change ¹⁴⁄₃₅ to a decimal. _____   173   **25**

(b) Change 7.325 to a fraction in
lowest terms. _____   173   **25**

4. Work with decimal fractions.

(a) What decimal part of 16.2 is 4.131?
(Round to three decimal places.) _____   173   **25**

(b) If 0.7 of a number is 12.67, find the
number. _____   173   **25**

If you are certain you can work all of these problems correctly, turn to
page 186 for a self-test. If you want help with any of these objectives or if
you cannot work one of the preview problems, turn to the page indicated. Super-students eager to learn everything in this unit will turn to
frame **1** and begin work there.

**ANSWERS TO
PREVIEW 3
PROBLEMS**

1. (a) 24.0762
     (b) 1.993

2. (a) 60.03
     (b) 716.856
     (c) 46.875
     (d) 4.16
     (e) 106.09

3. (a) 0.4
     (b) 7¹³⁄₄₀

4. (a) 0.255
     (b) 18.1

3   DECIMALS

© 1965 United Feature Syndicate, Inc.

1 Decimal Numbers

1 People who have trouble with math may not be able to add 2 and 2 in class or recognize equivalent fractions, but you can be certain they make no mistakes checking their change at the bank. Words never interfere with the really important things. Calculators are helpful, but it is important that you understand the process before you start pushing calculator keys. Otherwise you may get the wrong answer, lightning fast, with no clue that it is incorrect.

In Chapter 1 you learned that whole numbers are written in a place value system based on powers of ten. A number such as 237 is shown below in expanded form:

$$237$$
$$(2 \times 100) + (3 \times 10) + (7 \times 1)$$

Decimal This way of writing numbers can be extended to fractions. A **decimal** number is a fraction whose denominator is a power of 10. (Remember that a "power of 10" is simply a multiple of 10, that is, 10, 100, 1000, 10,000, and so on.) For example,

Decimal Form		Fraction Form
$0.6 =$	6 tenths	$= \dfrac{6}{10}$
$0.05 =$	5 hundredths	$= \dfrac{5}{100}$
$0.32 =$	32 hundredths	$= \dfrac{32}{100}$
$0.004 =$	4 thousandths	$= \dfrac{4}{1000}$
$0.267 =$	267 thousandths	$= \dfrac{267}{1000}$

We may also write the decimal number 0.267 in expanded form as shown here:

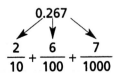

$$\frac{2}{10} + \frac{6}{100} + \frac{7}{1000}$$

Write the decimal number 0.526 in expanded form. Check your answer in **2**.

2 $0.526 = \dfrac{5}{10} + \dfrac{2}{100} + \dfrac{6}{1000}$

Decimal notation enables us to extend the idea of place value to numbers less than one. A decimal number often has both a whole number part and a fraction part. For example, the number 324.576 is shown in expanded form below:

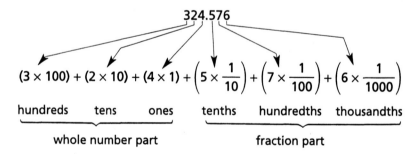

324.576

$$(3 \times 100) + (2 \times 10) + (4 \times 1) + \left(5 \times \dfrac{1}{10}\right) + \left(7 \times \dfrac{1}{100}\right) + \left(6 \times \dfrac{1}{1000}\right)$$

| hundreds | tens | ones | | tenths | hundredths | thousandths |

whole number part   fraction part

You are already familiar with this way of interpreting decimal numbers from working with money.

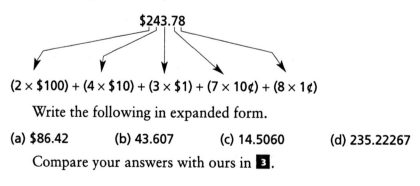

$243.78

$$(2 \times \$100) + (4 \times \$10) + (3 \times \$1) + (7 \times 10\cancel{c}) + (8 \times 1\cancel{c})$$

Write the following in expanded form.

(a) $86.42   (b) 43.607   (c) 14.5060   (d) 235.22267

Compare your answers with ours in **3**.

3 (a) $\$86.42 = (8 \times \$10) + (6 \times \$1) + \left(4 \times \$\frac{1}{10}\right) + \left(2 \times \$\frac{1}{100}\right)$

$= \$80 + \$6 + \$\frac{4}{10} + \$\frac{2}{100}$

(b) $43.607 = (4 \times 10) + (3 \times 1) + \left(6 \times \frac{1}{10}\right) + \left(0 \times \frac{1}{100}\right) + \left(7 \times \frac{1}{1000}\right)$

$= 40 + 3 + \frac{6}{10} + \frac{0}{100} + \frac{7}{1000}$

(c) $14.5060 = (1 \times 10) + (4 \times 1) + \frac{5}{10} + \frac{0}{100} + \frac{6}{1000} + \frac{0}{10000}$

(d) $235.22267 = (2 \times 100) + (3 \times 10) + (5 \times 1) + \frac{2}{10} + \frac{2}{100} + \frac{2}{1000}$

$+ \frac{6}{10000} + \frac{7}{100000}$

Notice that the denominators in the decimal fractions increase by a factor of 10. For example,

3247.8956

3 ×1000	Thousands	3000.0006	Ten-thousandths	6 × 0.0001
2 ×100	Hundreds	200.005	Thousandths	5 × 0.001
4 ×10	Tens	40.09	Hundredths	9 × 0.01
7 ×1	Ones	7.8	Tenths	8 × 0.1

⟵ —— Each row changes by a factor of ten. —— ⟶

$1 \times 10 = \quad 10 \quad 0.01 \quad \times 10 = 0.1$

$10 \times 10 = \quad 100 \quad 0.001 \quad \times 10 = 0.01$

$100 \times 10 = \quad 1000 \quad 0.0001 \quad \times 10 = 0.001$

Digits In the decimal number 86.423 the digits 4, 2, and 3 are called **decimal digits**. The number 43.6708 has four decimal digits. All digits to the right of the decimal point, those that name the fractional part of the number, are decimal digits.

How many decimal digits are included in the numeral 324.0576? Count them: then turn to **4**.

4 The number 324.0576 has four decimal digits: 0, 5, 7, and 6 (the digits to the right of the decimal point).

LEARNING HELP ▷ The decimal point is simply a way of separating the whole number part from the fraction part; it is a place marker. In whole numbers the decimal point usually is not written, but its location should be clear to you.

For a whole number the decimal point is understood to be after the right-end digit.

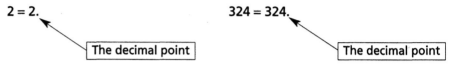

2 = 2.   324 = 324.

The decimal point   The decimal point

Very often additional zeros are annexed to decimal numbers without changing the value of the original number. For example,

8.5 = 8.50 = 8.5000   (and so on)
6 = 6. = 6.0 = 6.00   (and so on)

The value of the number is not changed, but the additional zeros may be useful, as we shall see.

LEARNING HELP ▷ The decimal number .6 is often written 0.6. The zero added on the left is used to call attention to the decimal point. It is easy to mistake .6 for 6, but the decimal point in 0.6 cannot be overlooked.

Add the following decimal numbers:

$$0.2 + 0.5 = \frac{2}{10} + \frac{5}{10} = \underline{\hspace{1cm}}$$

Try it, using what you know about adding fractions. Then turn to **5**.

2   Adding and Subtracting Decimal Numbers

5 $\quad 0.2 + 0.5 = \dfrac{2}{10} + \dfrac{5}{10} = \dfrac{2+5}{10} = \dfrac{7}{10} = 0.7$

We could skip the second step and say simply: 0.2 + 0.5 = 0.7.

Because decimal numbers represent fractions with denominators equal to powers of ten, addition is very simple.

$$2.34 \longrightarrow 2 + \frac{3}{10} + \frac{4}{100}$$

$$+\, 5.23 \longrightarrow \frac{5 + \dfrac{2}{10} + \dfrac{3}{100}}{7 + \dfrac{5}{10} + \dfrac{7}{100}} = 7.57$$

Add, using the expanded form shown above:

$$
\begin{array}{r}
1.45 \\
+\,3.42 \\
\end{array}
$$

Check your answer in **6**.

6

$$1.45 \longrightarrow 1 + \frac{4}{10} + \frac{5}{100}$$

$$+\,3.42 \longrightarrow \frac{3 + \dfrac{4}{10} + \dfrac{2}{100}}{4 + \dfrac{8}{10} + \dfrac{7}{100}} = 4.87$$

Of course we need not use expanded form in order to add decimal numbers. As with whole numbers, we may arrange the digits in vertical columns and add directly.

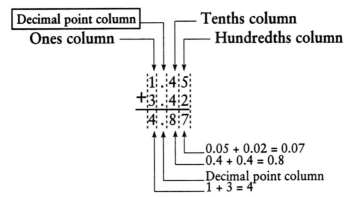

Digits of the same power of ten are placed in the same vertical column. Decimal points are always lined up vertically.

If one of the addends is written with fewer decimal digits than the other, annex as many zeros as needed to write both addends with the same number of decimal digits.

$$
\begin{array}{cccc}
2.345 & & & 2.345 \\
+\,1.5 & \text{becomes} & & +\,1.500 \\
\end{array}
$$

Except for the preliminary step of lining up decimal points, addition of decimal numbers is exactly the same process as addition of whole numbers.

Add the following decimal numbers.

(a) $\$4.02 + \$3.67 =$ _____   (b) $13.2 + 1.57 =$ _____

(c) $23.007 + 1.12 =$ _____   (d) $14.6 + 1.2 + 3.15 =$ _____

(e) $5.7 + 3.4 =$ _____   (f) $42.768 + 9.37 =$ _____

Arrange each sum vertically, placing the decimal points in the same vertical column. Then add as with whole numbers. Check your work in **7**.

7

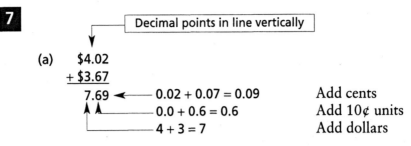

(a) $4.02
      + $3.67

      7.69 ◄——— 0.02 + 0.07 = 0.09   Add cents
            0.0 + 0.6 = 0.6   Add 10¢ units
            4 + 3 = 7   Add dollars

As a check, notice that the sum is roughly $4 + $3 or $7, which agrees with the actual answer. Always check your answer by first estimating it, then comparing your estimate or rough guess with the final answer.

| Decimal points in line |

(b)   13.20 ◄——— Annex a zero to provide the same number of decimal
      + 1.57   digits as in the other addend.
      14.77

              | Place answer decimal point in the same vertical line. |

Check: 13 + 1 = 14 (which agrees roughly with the answer)

(c)   23.007   (d)   14.60
      + 1.120   1.20
      24.127   + 3.15
                            18.95

(e)   5̷.7
      + 3.4
      9.1
            0.7 + 0.4 = 1.1; write 0.1, carry 1.
            Carried 1 + 5 + 3 = 9

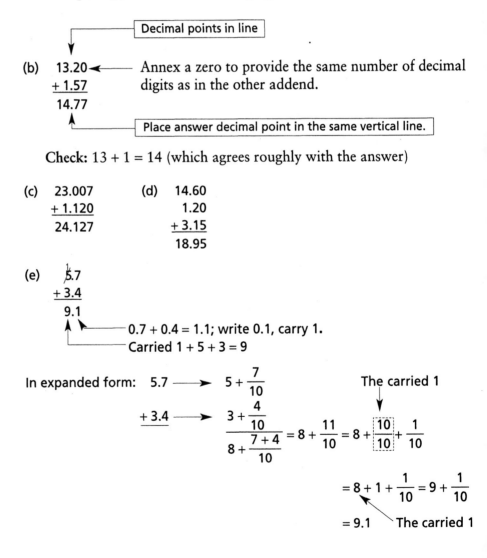

In expanded form: 5.7 ——► $5 + \dfrac{7}{10}$ The carried 1

                  + 3.4 ——► $3 + \dfrac{4}{10}$

                  $8 + \dfrac{7+4}{10} = 8 + \dfrac{11}{10} = 8 + \dfrac{10}{10} + \dfrac{1}{10}$

                  $= 8 + 1 + \dfrac{1}{10} = 9 + \dfrac{1}{10}$

                  $= 9.1$ The carried 1

(f)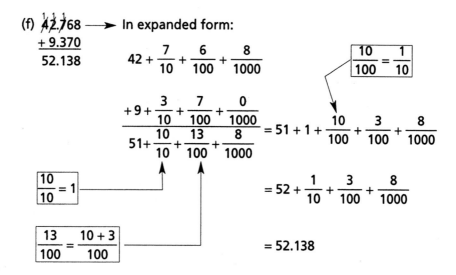

CAUTION ▷ You must line up the decimal points carefully to be certain of getting a correct answer.

Subtraction is equally simple if you line up decimal points carefully and attach any needed zeros before you begin work.

Example 1:

$437.56 − $41

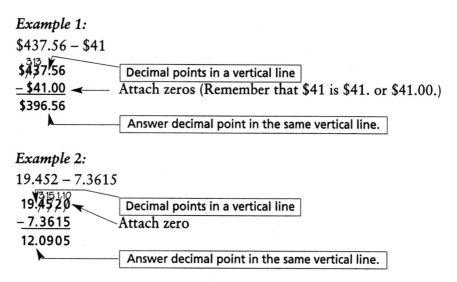

Try these problems to test yourself on subtraction of decimal numbers.

(a) $37.66 − $14.57 = _____   (b) 248.3 − 135.921 = _____
(c) 6.4701 − 3.2 = _____   (d) 7.304 − 2.59 = _____

Work carefully. The answers are in **8**.

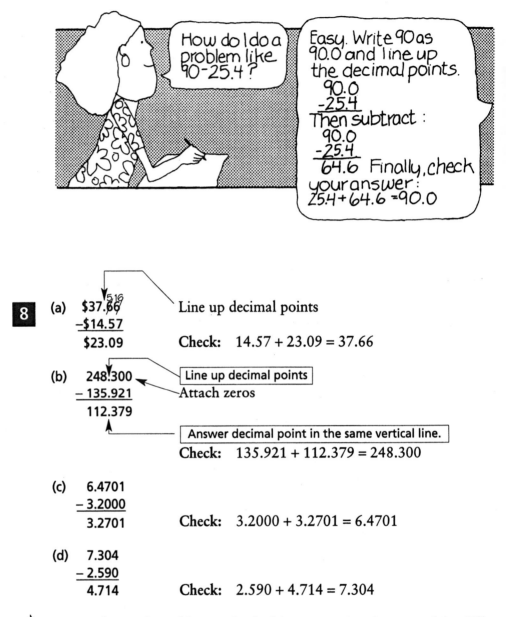

8 (a) $37.66 5 16
      −$14.57   Line up decimal points
      $23.09   **Check:** 14.57 + 23.09 = 37.66

(b) 248.300   Line up decimal points
    − 135.921   Attach zeros
    112.379

                        Answer decimal point in the same vertical line.
                        Check: 135.921 + 112.379 = 248.300

(c) 6.4701
    − 3.2000
      3.2701   **Check:** 3.2000 + 3.2701 = 6.4701

(d) 7.304
    − 2.590
      4.714   **Check:** 2.590 + 4.714 = 7.304

CAUTION Notice that each problem is checked by comparing the sum of the difference (answer) and subtrahend (number subtracted) with the minuend. Avoid careless mistakes by always checking your answer.

Now, for a set of practice problems on addition and subtraction of decimal numbers, turn to **9**.

9 *Problem Set 3-1:* Adding and Subtracting Decimal Numbers

A. Add or subtract as shown:

1. $0.5 + 0.3$	2. $0.7 + 0.9$
3. $0.9 + 0.6$	4. $0.4 + 0.2$
5. $0.1 + 0.8$	6. $0.5 + 0.7$
7. $0.8 + 0.8$	8. $0.3 + 0.4$
9. $0.8 + 0.9$	10. $0.8 - 0.7$
11. $0.9 - 0.2$	12. $5.6 - 2.3$
13. $0.9 - 0.4$	14. $4.9 - 2.6$
15. $2.9 - 1.1$	16. $1.7 - 0.3$
17. $3.7 - 0.4$	18. $8.7 - 3.5$
19. $0.7 + 0.9 + 0.3$	20. $5.2 + 1.7 + 3.0$
21. $2.8 + 0.9 + 1.1$	22. $0.5 + 0.8 + 0.1$
23. $2.6 + 4.5 + 1.9$	24. $8.3 + 2.6 + 7.2$
25. $1.4 + 3.6 + 0.5$	26. $0.3 + 0.6 + 0.5$
27. $8.8 + 3.4 + 5.3$	28. $3.3 - 1.7$
29. $9.3 - 2.6$	30. $7.2 - 6.6$
31. $4.2 - 0.6$	32. $7.1 - 5.8$
33. $5.7 - 3.9$	34. $8.5 - 5.9$
35. $3.0 - 0.4$	36. $1.1 - 0.7$

B. Add or subtract as shown:

1. $14.21 + 6.8$	2. $\$2.83 + \$12.19$
3. $0.687 + 0.93$	4. $3.76 + 23.43$
5. $\$7.04 + \$23.56$	6. $5.702 + 0.784$
7. $75.6 + 2.57$	8. $\$52.37 + \$98.74$
9. $0.096 + 5.82$	10. $507.18 + 321.42$
11. $4.0983 + 12.1036$	12. $623.09 + 408.19$

13. 45.6725 + 18.0588   **14.** 212.78 + 25.46

15. 70.3042 + 58.0643   **16.** 37 + .09 + 3.5 + 4.605

17. $14.75 + $9.08 + $3.76   **18.** 0.721 + 48.06 + 22 + 0.09

19. $52.19 + $17.43 + $38.75   **20.** 24.17 − 4.8

21. $33.40 − $18.04   **22.** 54.5 − 3.16

23. 7.83 − 6.79   **24.** $11.36 − $7.50

25. 75.08 − 32.75   **26.** 10.05 − 3.42

27. $20.00 − $13.48   **28.** 14.22 − 7.8

29. $40 − $3.82   **30.** 30 − 7.984

31. 3.892 − 0.995   **32.** $65 − $47.35

33. 13 − 6.04   **34.** 1.0487 − 0.6728

C. Calculate:

  1. 148.002 + 3.459   **2.** 632.9 − 30.246

  3. 68.708 + 27.18   **4.** 517.03 − 425.88

  5. 7.865 + 308.9   **6.** 23.745 − 9.06745

  7. 0.9437 + 15.0988   **8.** 4068.4 − 32.9067

  9. 8.939 + 10.072   **10.** 9.77803 − 6.42829

D. **Brain Boosters**

1. On a recent shopping spree at the Happy Peanut Health Food Store you bought the following:

1 qt celery juice	$3.75
1 jar honey	4.18
Granny's Granola	5.36
sunflower seeds	2.59
virgin olive oil	8.98

How much change should you receive from a $50 bill?

2. At the start of a long trip your mileage meter read 18327.4, and at the end of the trip it read 23015.2. How far did you travel?

3. **Business** The Zippy Delivery Service truck logged 124.7 delivery miles Monday, 209.1 miles Tuesday, 279.4 miles Wednesday, 185.4

miles Thursday, and 313.9 miles Friday. What was the total mileage for the week?

4. The 400-meter race run in the Olympic games is 437.444 yards long. What is the difference between this distance and one-quarter of a mile (440 yards)?

5. **Machinist** A certain machine part is 2.345 inches thick. What is its thickness after 0.078 inch is ground off?

6. **Finance** Can you balance a checkbook? At the start of a shopping spree your balance was $1472.33. While shopping you write checks for $212.57, $18.95, $40, $7.48, and $523.98. What is your new balance?

7. **Real Estate** Isabelle from Radis Realty needs to find the perimeter (distance around) the field shown. All distances are in meters. Help her with the calculation.

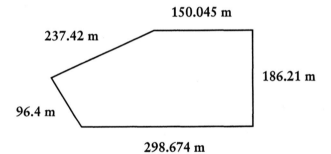

150.045 m

237.42 m

186.21 m

96.4 m

298.674 m

8. **Auto Mechanic** Otto, the mechanic, finds that a heated piston measures 8.586 cm in diameter. When cold it measures 8.573 cm in diameter. How much does it expand when heated?

9. **Sales** What is the actual cost of the following car?

Sticker price	$17,745.00
Dealer preparation	206.50
Leather interior	875.40
Antilock brakes	939.00
CD player	419.95
Moonroof	735.50
Tax and license	618.62
Less trade-in	1780.00

E. Calculator Problems

Solve using a calculator.

1. 23770.66 + 34.07 + 1664.82

2. $467.33 + $235.05 + $46.22 + $110.46 + $54.98

3. 23,456.1 + 44,005.6 + 11,873.9 + 2,340.23

4. 123,456.7 + 253.46 + 82.59 + 66,765.3 + 344,700.08

5. $4.76 + $2.98 + $21.66 + $4.92 + $13.07 + $4.50 + $9.34

6. 25.0982 + 1.00258 + 61.06668 + 7.770771 + 34.5 + 21.2280

7. 22,222.2 + 2.222 + 2222.22 + 220.22 + 22.220

8. 56,678.34 – 9,870.67

9. 423,780.8 – 6,670.9

10. 1120.34 – 998.664

11. 356.678 – 21.99 + 45.239 – 101.7

12. $56.37 – $127.20 + $495.03 + $12.46 – $64.55

13. Finance   At the start of the month Enrique's checking account had a balance of $846.71. During the month he wrote checks for the following amounts: $34.98, $18.37, $74.76, $93.54, $236.45, $121.35, and $29.33. He deposited $408.37 during the month. What is his balance in the account?

14. One minute is defined as exactly 60 seconds in sun time. One minute as measured by movement of the earth with respect to the stars is 59.836174 seconds. Find the difference between sun time and star time for one minute.

The answers to these problems are in the Appendix. When you have had the practice you need either return to the preview on page 143 or continue in frame **10** with the study of multiplication and division of decimal numbers.

3   Multiplying and Dividing Decimal Numbers

10   A decimal number is a fraction having a power of ten as denominator. Multiplication of decimals should therefore be no more difficult than the multiplication of fractions. Try this problem:

$0.5 \times 0.3 =$ _____

Write out the two numbers as fractions and multiply, and then choose an answer.

(a) 15 Go to **11**.
(b) 1.5 Go to **12**.
(c) 0.15 Go to **13**.

11 You answered that $0.5 \times 0.3 = 15$ and that is incorrect. A wise first step would be to guess at the answer. Both 0.5 and 0.3 are less than 1; therefore their product is also less than 1, and 15 is not a reasonable answer.

Try calculating the sum this way:

$$0.5 = \frac{5}{10} \qquad 0.3 = \frac{3}{10}$$

$$0.5 \times 0.3 = \frac{5}{10} \times \frac{3}{10}$$

Complete this multiplication, and then return to **10** and choose a better answer.

12 Your answer is incorrect. Don't get discouraged; we'll never tell.

The first step is to guess: both 0.5 and 0.3 are less than 1; therefore their product will be less than 1. Next, convert the decimals to fractions with 10 as denominator.

$$0.5 = \frac{5}{10} \qquad 0.3 = \frac{3}{10}$$

Finally, multiply.

$$0.5 \times 0.3 = \frac{5}{10} \times \frac{3}{10} = \underline{\hspace{2cm}}$$

Complete this multiplication, and then return to **10** and choose a better answer.

13 Excellent. Notice that both 0.5 and 0.3 are less than 1; therefore their product will be less than 1. This provides a rough guess at the answer.

$$0.5 = \frac{5}{10} \qquad 0.3 = \frac{3}{10}$$

$$0.5 \times 0.3 = \frac{5}{10} \times \frac{3}{10} = \frac{5 \times 3}{10 \times 10} = \frac{15}{100} = 0.15$$

Of course it would be very, very clumsy and time consuming to calculate every decimal multiplication in this way. We need a simpler method. Here is the procedure most often used:

Step 1   Multiply the two decimal numbers as if they were whole numbers. Pay no attention to the decimal points.

Step 2   The sum of the decimal digits in the factors will give you the number of decimal digits in the product.

Let's apply this procedure to the problem 3.2×0.41.

Step 1   Multiply without regard to the decimal points.

$$\begin{array}{r} 32 \\ \times\,41 \\ \hline 1312 \end{array}$$

Step 2   Count the number of decimal digits in the factors.

3.2 has *one* decimal digit (2).
0.41 has *two* decimal digits (4 and 1).

The total number of decimal digits in the two factors is 3. The product will have *three* decimal digits. Count over *three* digits to the left in the product.

1.312

Three decimal digits

Check: 3.2×0.41 is roughly $3 \times \frac{1}{2}$ or about $1\frac{1}{2}$. The answer 1.3 agrees with our rough guess.

Try these simple decimal multiplications.

(a) $0.5 \times 0.5 =$ _____

(b) $0.1 \times 0.1 =$ _____

(c) $10 \times 0.6 =$ _____

(d) $2 \times 0.4 =$ _____

(e) $1 \times 0.1 =$ _____

(f) $2 \times 0.003 =$ _____

(g) $0.01 \times 0.02 =$ _____

(h) $0.04 \times 0.005 =$ _____

Follow the steps outlined above. Count decimal digits carefully. Check your answers in **14**.

14 (a) $0.5 \times 0.5 = $ _____

First, multiply $5 \times 5 = 25$. Second, count decimal digits:

0.5 (one decimal digit) × 0.5 (one decimal digit) = a total of *two* decimal digits

Count over *two* decimal digits from the right: .25
The product is 0.25.

Check: Both factors (0.5) are less than 1; therefore their product will be less than 1, and 0.25 seems reasonable.

(b) 0.1×0.1 $1 \times 1 = 1$

Count over *two* decimal digits from the right. Since there are not two decimal digits in the product, attach a few zeros on the left.

1 ——▸ 0.01

(Two decimal digits

So $0.1 \times 0.1 = 0.01$

Check: $\dfrac{1}{10} \times \dfrac{1}{10} = \dfrac{1}{100}$ OK.

(c) 10×0.6 $10 \times 6 = 60$

Count over *one* decimal digit from the right (6.0) so that $10 \times 0.6 = 6.0$. Notice that multiplication by 10 simply shifts the decimal place one digit to the right.

$10 \times 6.2 = 62$
$10 \times 0.075 = 0.75$
$10 \times 8.123 = 81.23$ (and so on)

(d) 2×0.4 $2 \times 4 = 8$

Count over *one* decimal digit.   .8 $2 \times 0.4 = 0.8$

(e) 1×0.1 $1 \times 1 = 1$

Count over *one* decimal digit.   .1 $1 \times 0.1 = 0.1$

(f) 2×0.003 $2 \times 3 = 6$

Count over *three* decimal digits.   0.006 $2 \times 0.003 = 0.006$

(g) 0.01×0.02 $1 \times 2 = 2$

0.01 (two decimal digits) $\times$ 0.02 (two decimal digits) = total of four decimal digits
Count over *four* decimal digits.   0.0002 $0.01 \times 0.02 = 0.0002$

(h) 0.04×0.005 $4 \times 5 = 20$

0.04 (two decimal digits) $\times$ 0.005 (three decimal digits) = total of five decimal digits
Count over *five* decimal digits.   0.00020 $0.04 \times 0.005 = 0.0002$

CAUTION ▷
- Do not try to do this entire process mentally until you are certain you will not misplace zeros.

- Always estimate before you begin the arithmetic, and finally check your answer against your estimate. (Estimating and checking are particularly important when you use a calculator to do the work.)

Multiplication of larger decimal numbers is performed in exactly the same manner. Try these:

(a) $4.302 \times 12.05 =$ _____

(b) $6.715 \times 2.002 =$ _____

(c) $3.144 \times 0.00125 =$ _____

Look in **15** for the answers.

How to Name Decimal Numbers

The decimal number 3,254,935.4728 should be interpreted as:

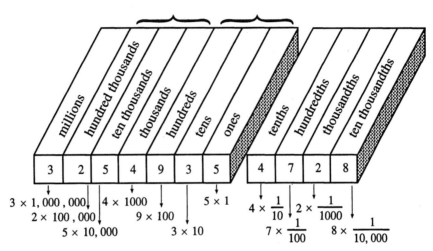

$$3 \times 1,000,000$$
$$2 \times 100,000$$
$$5 \times 10,000$$
$$4 \times 1000$$
$$9 \times 100$$
$$3 \times 10$$
$$5 \times 1$$
$$4 \times \frac{1}{10}$$
$$7 \times \frac{1}{100}$$
$$2 \times \frac{1}{1000}$$
$$8 \times \frac{1}{10,000}$$

It may be read "three million, two hundred fifty-four thousand, nine hundred thirty-five, and four thousand one hundred twenty-eight ten thousandths."

Notice that the decimal point is read "and."

It is useful to recognize that in this number the digit 8 represents 8 ten-thousandths or $\frac{8}{10,000}$, and the digit 7 represents 7 hundredths or $\frac{7}{100}$.

This number is often read more simply as "three million, two hundred fifty-four thousand, nine hundred thirty-five, *point* four, seven, two, eight." This way of reading the number is easiest to write, to say, and to understand.

---

15 (a) **Guess:** 4×12 is 48. The product will be about 48.

$$\begin{array}{r} 4302 \\ \times\ 1205 \\ \hline 5183910 \end{array}$$

(If you cannot do this multiplication correctly, turn to page 36 in Chapter 1 for help with the multiplication of whole numbers.)

The factors contain a total of five decimal digits (three in 4.302 and two in 12.05). Count over five decimal digits from the right in the product

51.83910

so that $4.302 \times 12.05 = 51.8391$.

Check: The answer, 51.8391, is approximately equal to the guess, 48.

(b) **Guess:** 6.7×2 is about 7×2 or 14.

$$
\begin{array}{r}
6715 \\
\times\, 2002 \\
\hline
13.443430
\end{array}
$$

6715 ◄——— 6.715 has *three* decimal digits
× 2002 ◄——— 2.002 has *three* decimal digits
13.443430 ◄——— A total of *six* decimal digits

Six decimal digits $6.715 \times 2.002 = 13.44343$

Check: The answer and the guess are approximately equal.

(c) **Guess:** 3×0.001 is about 0.003.

$$
\begin{array}{r}
3144 \\
\times\, 125 \\
\hline
.00393000
\end{array}
$$

3144 ◄——— 3.144 has *three* decimal digits
× 125 ◄——— 0.00125 has *five* decimal digits
.00393000 ◄——— A total of *eight* decimal digits

Eight decimal digits $3.144 \times 0.00125 = 0.00393$

Check: The answer and the guess are approximately equal.
Now go to **16** for a look at division of decimal numbers.

16 **Division**   Division of decimal numbers is very similar to division of whole numbers. For example, $6.8 \div 1.7$ can be written:

$$\frac{6.8}{1.7}$$

and if we multiply the top and bottom of the fraction by 10,

$$\frac{6.8}{1.7} = \frac{6.8 \times 10}{1.7 \times 10} = \frac{68}{17}$$

Now treat $^{68}\!/_{17}$ as a normal whole number division:

$$\frac{68}{17} = 68 \div 17 = 4$$

Therefore, $6.8 \div 1.7 = 4$.   **Check:** $1.7 \times 4 = 6.8$.
Rather than take the trouble to write the division as a fraction, we may use a shortcut.

Step 1	**Example**
Write the divisor and dividend in standard long division form.	$6.8 \div 1.7 = ?$ $1.7\overline{)6.8}$

Step 2

Shift the decimal point in the divisor to the
right so as to make the divisor a whole number.   1.7)‾‾‾

Step 3

Shift the decimal point in the dividend *the
same amount*. (Add zeros if necessary.)   1.7)‾6.8.‾

Step 4

Place the decimal point in the answer space
directly above the new decimal position in
the dividend.   17)‾68.‾

Step 5

Complete the division exactly as you would   4.
with whole numbers. The decimal points in   17)‾68.‾
divisor and dividend may now be ignored.   68

$$6.8 \div 1.7 = 4$$

Notice in steps 2 and 3 we have simply multiplied both divisor and dividend by 10.

Repeat the process outlined above with this division:

$$1.38 \div 2.3$$

Work carefully, then compare your work with ours in **17**.

17   Let's do it step by step.

2.3)‾1.38‾

2.3.)‾1.3.8‾

23.)‾13.8‾

Shift the decimal point one digit to
the right so that the divisor becomes a whole number. Then shift
the decimal point in the dividend
the *same* number of digits. 2.3 becomes 23.

1.38 becomes 13.8. This is the same
as multiplying both numbers by 10.

Place the answer decimal point
directly above the decimal point in
the dividend.

$$\begin{array}{r} .6 \\ 23.\overline{)13.8} \\ 13\ 8 \end{array}$$

Divide as you would with whole numbers.

$$1.38 \div 2.3 = 0.6$$

Check: $2.3 \times 0.6 = 1.38$

Always remember to check your answer.
How would you do this one?

$$2.6 \div 0.052 = \underline{\hspace{1.5cm}}$$

Look in **18** for the solution after you have tried it.

18   .052.$\overline{)2.6}$

To shift the decimal place three digits in the dividend, we must attach several zeros to its right.

.052.$\overline{)2.600.}$

$$\begin{array}{r} 50. \\ 52.\overline{)2600.} \\ \underline{260\downarrow} \\ 0 \\ \underline{0} \end{array}$$

Then place the decimal point in the answer space above that in the dividend, and divide as with whole numbers.

$$2.6 \div 0.052 = 50$$

Check: $0.052 \times 50 = 2.6$

Shifting the decimal point three digits and attaching zeros to the right of the decimal point in this way is equivalent to multiplying both divisor and dividend by 1000.
Try these problems:

(a) $3.5 \div 0.001 = \underline{\hspace{1.5cm}}$ (b) $9 \div 0.02 = \underline{\hspace{1.5cm}}$

(c) $0.365 \div 18.25 = \underline{\hspace{1.5cm}}$ (d) $8.8 \div 3.2 = \underline{\hspace{1.5cm}}$

The answers are in **19**.

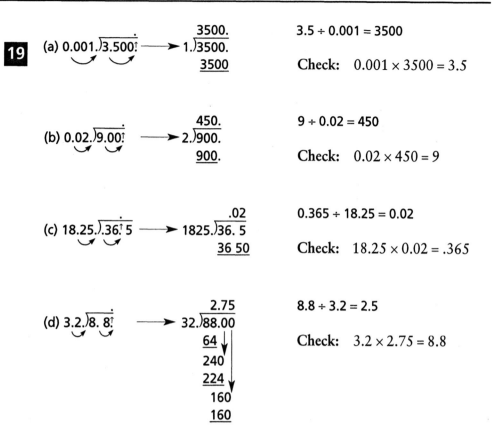

19   (a) $0.001.\overline{)3.500}$ ⟶ $1.\overline{)3500.}$

$\begin{array}{r} 3500. \\ \hline \underline{3500} \end{array}$

$3.5 \div 0.001 = 3500$

Check: $0.001 \times 3500 = 3.5$

(b) $0.02.\overline{)9.00}$ ⟶ $2.\overline{)900.}$

$\begin{array}{r} 450. \\ \hline \underline{900.} \end{array}$

$9 \div 0.02 = 450$

Check: $0.02 \times 450 = 9$

(c) $18.25.\overline{)36.5}$ ⟶ $1825.\overline{)36.5}$

$\begin{array}{r} .02 \\ \hline \underline{36\ 50} \end{array}$

$0.365 \div 18.25 = 0.02$

Check: $18.25 \times 0.02 = .365$

(d) $3.2.\overline{)8.8}$ ⟶ $32.\overline{)88.00}$

$\begin{array}{r} 2.75 \\ \underline{64} \\ 240 \\ \underline{224} \\ 160 \\ \underline{160} \end{array}$

$8.8 \div 3.2 = 2.5$

Check: $3.2 \times 2.75 = 8.8$

If the dividend is not exactly divisible by the divisor, we must either stop the process after some preset number of decimal places in the answer or we must round the answer. We do not generally indicate a remainder in decimal division.

Turn to **20** for some rules for rounding.

20   **Rounding**   **Rounding** is a process of approximating a number. To round a number means to find another number roughly equal to the given number but expressed less precisely. For example,

$432.57 = 400 rounded to the nearest hundred dollars
          $= 430 rounded to the nearest ten dollars
          $= 433 rounded to the nearest dollar

$1.376521 = 1.377$ rounded to three decimal digits
          $= 1.4$ rounded to the nearest tenth
          $= 1$ rounded to the nearest whole number

There are exactly 5280 feet in 1 mile. To the nearest thousand feet how many feet are in one mile? To the nearest hundred feet?

Check your answers in .

 5280 ft = 5000 ft rounded to the nearest thousand feet
            (In other words 5280 is closer to 5000 than to 4000 or 6000.)
5280 ft = 5300 ft rounded to the nearest hundred feet
            (In other words 5280 is closer to 5300 than to 5200 or 5400.)

You may want to review the rounding of whole numbers on page 64.

To round decimal numbers, follow these simple steps:

Step 1	**Example**
Determine the number of digits or the place to which the number is to be rounded. Mark it with a ‸.	Round 3.462 to one decimal place: 3.4‸62

Step 2

If the digit to the right of the mark is less than 5, replace all digits to the right of the mark by zeros. If the zeros are decimal digits, you may discard them.	2.8‸32 becomes 2.800 or 2.8

Step 3

If the digit to the right of the mark is equal to or larger than 5, increase the digit to the left by 1.	3.4‸62 becomes 3.5

Try applying this rounding procedure to these problems.

(a) Round 74.238 to two decimal places.

(b) Round 8.043 to two decimal places.

(c) Round 21.3754 to the nearest thousandth.

(d) Round 6.07 to the nearest tenth.

Follow the rules, then check your work in **22**.

22   (a) 74.238 = 74.24 to two decimal places
(Write 74.23‚8 and note that 8 is larger than 5; so increase the 3 to 4.)

(b) 8.043 = 8.04 to two decimal places
(Write 8.04‚3 and note that 3 is less than 5; so drop it.)

(c) 21.3754 = 21.375 to the nearest thousandth
(Write 21.375‚4 and note that the digit to the right of the mark is 4; so drop it.)

(d) 6.07 = 6.1 to the nearest tenth
(Write 6.0‚7 and note that 7 is greater than 5; so increase the 0 to 1.)

In the following problem, divide as shown and round your answer to two decimal places.

6.84 ÷ 32.7 = _____

Careful now. Check your work in **23**.

Multiplying and Dividing by Powers of Ten

Many practical problems involve multiplying or dividing by 10, 100, or 1000. You will find it very useful to be able to multiply and divide by powers of 10 quickly and without using paper, pencil, or a calculator. To begin, you may want to review the section on multiplying a whole number by a multiple of 10 on page 42.

The following rules are useful when you must multiply a decimal number by a multiple of 10.

1. To multiply a decimal number by 10, 100, 1000, or a larger multiple of 10, move the decimal point as many places to the **right** as there are zeros in the multiplier.

$$8.24 \times 10 = 82.4$$ Shift the decimal point 1 place to the right.

$$3.458 \times 100 = 345.8$$ Shift the decimal point 2 places to the right.

- You may need to attach additional zeros before moving the decimal point.

$$2.4 \times 1000 = 2.400 \times 1000 = 2400.\text{ or } 2400$$

> Attach two zeros, then
> shift the decimal point
> 3 places to the right.

2. To divide a whole number of a decimal number by 10, 100, 1000, or a larger multiple of 10, move the decimal point as many places to the **left** as there are zeros in the divisor.

$$12.4 \div 10 = 1.24$$ Shift the decimal point 1 place to the left.

$$37.8 \div 100 = .378 = 0.378$$ Shift the decimal point 2 places to the left.

- You may need to attach additional zeros before moving the decimal point.

$$3.4 \div 1000 = 003.4 \div 1000 = .0034 = 0.0034$$

> Attach 2 zeros on the left;
> then shift the decimal
> point 3 places to the left.

- With a whole number the decimal point is usually not written, and you must remember that it is understood to be after the units digit.

$$4 = 4.$$

$$4 \div 100 = 4. \div 100 = .04 = 0.04$$

> Move the decimal point
> 2 places to the left.

Here are a few problems for practice. Work quickly. No pencil and paper needed. Do them in your head.

1. 42.38×10	**2.** 529.237×100
3. 6.226×100	**4.** 1.7774×100
5. 2.4401×1000	**6.** 5.0037×1000

7. 0.2388 × 100

8. 0.04805 × 1000

9. 0.4 × 100

10. 1.8 × 1000

11. 24.66 ÷ 10

12. 70.347 ÷ 10

13. 237.882 ÷ 100

14. 5170.2921 ÷ 100

15. 4.57 ÷ 1000

16. 0.2792 ÷ 1000

17. 0.02 ÷ 100

18. 0.0037 × 100

19. 0.0045 ÷ 100

20. 0.00057 × 100

Check your answers in the Appendix.

---

23 32.7.)6.8ɾ4

```
        .209
327.)68.400
     65 4          2 × 327 = 654
     3 000
     2 943         9 × 327 = 2943
```

0.209 = 0.21 rounded to two decimal places

6.84 ÷ 32.7 ≈ 0.21 rounded to two decimal places.

Check: 32.7 × 0.21 = 6.867 which is approximately equal to 6.84. (The check will not be exact because we have rounded.)
  Go to for a set of practice problems on multiplication and division of decimal numbers.

24 *Problem Set 3-2:* Multiplying and Dividing Decimal Numbers

A. Multiply:

1. 0.01 × 0.001

2. 10 × 0.01

3. 10 × 2.15

4. 3 × 0.02

5. 0.04 × 0.2

6. 0.07 × 0.2

7. 0.3 × 0.3

8. 0.9 × 0.8

9. 1.2 × 0.7

10. 4.5 × 0.002

11. 0.005×0.012

12. 3.5×1.2

13. 6.41×0.23

14. 7.25×0.301

15. 16.2×0.031

16. $0.2 \times 0.3 \times 0.5$

17. $0.5 \times 1.2 \times 0.04$

18. $0.6 \times 0.6 \times 6.0$

19. $1.2 \times 1.23 \times 0.01$

20. $2.3 \times 1.5 \times 1.05$

21. $1.2 \times 10 \times 0.12$

22. 321.4×0.25

23. 0.234×0.005

24. 125×2.3

25. 5.224×0.00625

26. 0.1234×0.0075

27. 425.6×2.875

28. 0.0079×0.023

B. Divide:

1. $6.5 \div 0.005$

2. $3.78 \div 0.30$

3. $0.0405 \div 0.9$

4. $6.5 \div 0.5$

5. $0.378 \div 0.003$

6. $40.5 \div 0.09$

7. $3 \div 0.05$

8. $12 \div 0.006$

9. $10 \div 0.001$

10. $2.59 \div 70$

11. $1.2321 \div 0.111$

12. $44.22 \div 6.7$

13. $57.57 \div 0.0303$

14. $104.2 \div 0.0320$

15. $1.111 \div 10.1$

16. $0.0747 \div 0.0332$

C. Divide and round as indicated.

Round to two decimal digits:

1. $10 \div 3$

2. $1 \div 0.7$

3. $5 \div 6$

4. $0.07 \div 0.80$

5. $2.0 \div 0.19$

6. $2 \div 3$

7. $1 \div 4$

8. $1 \div 8$

9. $100 \div 9$

10. $20 \div 0.07$

11. $0.006 \div 0.04$

12. $0.8 \div 0.05$

Round to three decimal digits:

13. $10 \div 0.70$

14. $0.04 \div 1.71$

15. $0.09 \div 0.40$

16. $0.091 \div 0.0014$

17. $22.4 \div 6.47$

18. $3.41 \div 0.25$

19. $3.51 \div 0.92$

20. $6.001 \div 2.001$

21. $4.0 \div 0.007$

22. $123.321 \div 0.111$

D. Calculate as indicated:

1. $(0.3)^2$

2. $(0.03)^2$

3. $(0.003)^2$

4. $(0.3)^3$

5. $(1.2)^2$

6. $(1.2)^3$

7. $(0.1)^2$

8. $(0.01)^2$

9. $(0.03)^3$

Round to three decimal digits:

10. $\dfrac{1}{81}$

11. $\dfrac{1}{7}$

12. $\dfrac{0.23 \times 7.5}{0.23 + 7.5}$

13. $\dfrac{0.02 \times 3.2}{0.2 + 3.2}$

14. $\dfrac{0.065 - 0.042}{0.065 + 0.042}$

15. $\dfrac{0.03 \div 0.006}{0.03 + 0.006}$

E. Brain Boosters

1. Andy worked 37.4 hours at $6.25 per hour. How much money did he earn?

2. What is the cost of 12.3 gallons of gasoline at $1.879 per gallon?

3. **Business** A television set is advertised for $420. It can also be bought "on time" for 24 payments of $22.75 each. How much extra do you pay by purchasing it on the installment plan?

4. **Medical Technology** The Fit-4-U weight loss center calculates the Body Mass Index (BMI) for all prospective clients. It is found by multiplying their weight in pounds by 703 and dividing that product by the square of their height in inches. A person with a BMI of 30 or more is considered obese; a BMI over 25 is considered overweight. Calculate the BMI of a person 5 ft 6 in. tall weighing 165 pounds.

5. **Meteorology**   When the wind blows strongly the air temperature feels colder than the thermometer indicates. The wind-chill factor is found by multiplying the wind speed by 1.5 and subtracting that product from the actual temperature. Calculate the wind-chill factor on a day when the wind is flowing at 30 mph and the actual temperature is 45 degrees Fahrenheit.

6. There are three errors in these five problems. What are they?

   (a) $6 \div 0.02 = 300$

   (b) $12 \div 0.03 = 4$

   (c) $2.4 \times 0.3 = 0.72$

   (d) $100 \times 0.002 = 0.20$

   (e) $0.2 \times 0.2 = 0.4$

7. Lyn once said, "The day before yesterday I was 19, but next year I'll be 22." She always tells the truth. When did she say this?

8. Is this problem correct?

   $$\frac{6 + 6 - 0.6}{0.6} = 19$$

   Try substituting any other digit for 6. Notice anything interesting?

9. **Veterinary Medicine**   An adult dog will consume one ounce of food for every 2 pounds of body weight (his, not yours). (a) What will Tiger, my 25 lb terrier, eat per day? (b) How much will Tamo, my 94 lb German shepherd, eat? (c) If dry kibble costs $7.95 per 25 lb bag, what does it cost to feed these pets each week?

10. **Fire Protection**   One gallon of water weighs 8.34 lb. How much weight is added to a fire truck when its tank is filled with 750 gallons of water?

F. Calculator Problems
Round to three decimal places.

   1. 0.234×5.877

   2. 6.667×0.9098

   3. 23.55×0.0884

   4. 3.448×1.758

   5. $45.79 \div 68.689$

   6. $0.8237 \div 8.279$

   7. $14.993 \div 2.367$

   8. $0.0287 \div 0.00698$

Round to four decimal places.

   9. 238.456×0.007625

   10. $0.024562 \div 0.983983$

   11. $2.2223 \div 7.7774$

   12. 3.14159×78.356

13. $1 \div 9$ **14.** $3 \div 77$

15. 2.8283×0.7731 **16.** $0.0056729 \times 4842.375$

The answers to these problems are in the Appendix. When you have had the practice you need, either return to the preview on page 143 or continue in frame **25** with the study of decimal fractions.

DECIMAL FRACTIONS

25 Since decimal numbers are fractions, they may be used, as fractions are used, to represent a part of some quantity. For example, recall that

"$\dfrac{1}{2}$ of 8 equals 4" means $\dfrac{1}{2} \times 8 = 4$

and therefore,

"0.5 of 8 equals 4" means $0.5 \times 8 = 4$

The word *of* used in this way indicates multiplication.

Find 0.35 of 8.4.

(If you need a review of problems of this kind, turn to frame **51** in Chapter 2.) Go to **26** to check your answer to this problem.

26 $0.35 \times 8.4 = $ _____

$$
\begin{array}{r}
8.4 \\
\times\,0.35 \\
\hline
420 \\
252 \\
\hline
2.940
\end{array}
$$

8.4 ◄——— One decimal digit
× 0.35 ◄——— Two decimal digits
2.940 ◄——— A total of three decimal digits

Here is a second variety of problem:

What part of 16 is 4?

$$\square \times 16 = 4$$

$$\square = 4 \div 16 = \dfrac{1}{4}$$

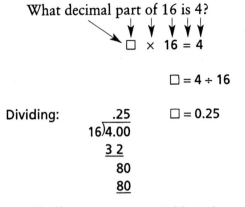

What decimal part of 16 is 4?

□ × 16 = 4

□ = 4 ÷ 16

Dividing: .25   □ = 0.25
        16)4.00
           3 2
           80
           80

Check: 0.25 × 16 = 4.00 or 4

What decimal part of 8 is 3? Solve this problem using the method illustrated above; then hop to **28**.

27   (a) □ × 5 = 13

          □ = 13 ÷ 5
          □ = 2.6

     Check: 2.6 × 5 = 13.0 or 13

          2.6
        5)13.0
          10
          30
          30

   (b) 0.8 × □ = 10

          □ = 10 ÷ 0.8

          □ = 12.5

     Check: 0.8 × 12.5 = 10.00 or 10

            1 2.5
       0.8.)10.0.0
              8
             2 0
             1 6
              4 0
              4 0

   (c) 2.35 × □ = 1.739
          □ = 1.739 ÷ 2.35
          □ = 0.74

     Check: 2.35 × 0.74 = 1.7390 or 1.739

              .74
      2.35.)1.73.90
            1 64 5
               9 40
               9 40

Terminating   To convert a number from fraction form to decimal form, simply divide as
Decimal   indicated in the problems above. If the division has no remainder, the decimal number is called a **terminating decimal**.

For example, $\dfrac{5}{8} =$ _____

```
     .625
8)5.000 ◄——— Attach as many zeros as needed.
   4 8
     20
     16
     40
     40
      0 ◄——— Zero remainder; hence the decimal terminates or ends.
```

$\dfrac{5}{8} = 0.625$

If a decimal does not terminate, you may round it to any desired number of decimal digits. Try this one.

$\dfrac{2}{13} =$ _____

Divide it out and round to three decimal digits. Check your work in **29**.

28   What decimal part of 8 is 3?

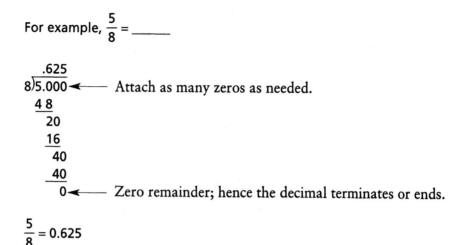

```
   □ × 8 = 3
                          .375
                       8)3.000
       □ = 3 ÷ 8          2 4
       □ = 0.375           60
                           56
                           40
Check:   0.375 × 8 = 3.000 or 3    40
```

Try these:

(a) What decimal part of 5 is 13?
(b) If 0.8 of a number is 10, find the number.
(c) If 2.35 of a number is 1.739, find the number.

Look in **27** for the answers.

29

$$13\overline{)2.0000} \quad \frac{.1538}{}$$

$$\frac{2}{13} = 0.154 \text{ rounded to three decimal digits}$$

```
      .1538
13)2.0000
    1 3
     70
     65
     50
     39
    110
    104
      6
```

Convert the following fractions to decimal form, and round to two decimal digits.

(a) $\frac{2}{3} =$ (b) $\frac{5}{6} =$ (c) $\frac{17}{7} =$ (d) $\frac{7}{16} =$

Our work is in **30** .

30

(a)
```
      .666
3)2.000
   1 8
    20
    18
    20
    18
     2
```

$\frac{2}{3} = 0.67$ rounded to two decimal digits

LEARNING HELP ▷ Notice that in order to round to two decimal digits, we must carry the division out to at least three decimal digits.

(b)
```
     .833
  6)5.000
     4 8
      20
      18
      20
      18
       2
```
$\dfrac{5}{6} = 0.83$ rounded to two decimal digits

(c)
```
    2.428
  7)17.000
    14
     3 0
     2 8
      20
      14
      60
      56
       4
```
$\dfrac{17}{7} = 2.43$ rounded to two decimal digits

(d)
```
     .437
  16)7.000
     6 4
      60
      48
      120
      112
        8
```
$\dfrac{7}{16} = 0.44$ rounded to two decimal digits

Repeating Decimal Decimal numbers that do not terminate repeat a sequence of digits. Such decimals are called **repeating decimals.** For example,

$$\frac{1}{3} = 0.333 \cdots$$

The three dots are read "and so on" and indicate that the digit 3 continues without end. Similarly,

$$\frac{2}{3} = 0.6666 \cdots$$

and $\frac{3}{11}$ is

$$
\begin{array}{r}
.2727 \\
11\overline{)3.0000} \\
\underline{2\,2} \\
80 \\
\underline{77} \\
30 \\
\underline{22} \\
80 \\
\underline{77} \\
3
\end{array}
$$

or $\frac{3}{11} = 0.272727\cdots$

Notice the remainder 3 is equal to the original dividend. This indicates that the decimal quotient repeats itself.

We may use a shorthand notation to show that the decimal repeats.

$$\frac{1}{3} = 0.\overline{3} \qquad\qquad \frac{2}{3} = 0.\overline{6}$$

The bar means that the digits under the bar repeat endlessly.

$$\frac{3}{11} = 0.\overline{27} \text{ means } 0.272727\cdots$$

$$3\frac{1}{27} = 3.\overline{037} \text{ means } 3.037037037\cdots$$

Write $\frac{4}{33}$ as a repeating decimal using the "bar" notation. Check your answer in **31**.

Measuring Concentrations

The daily newspaper uses a very useful sort of measurement unit when it reports that the concentration of sulfur dioxide is 0.2 ppm (parts per million) in Detroit air during a smog alert. If we need to specify the concentration of very small relative amounts of material (as in air or water pollution or medicine dosages), we can write it as the ratio of the amount of substance added to the total amount of material. For example, adding 1 pound of salt to 1000 pounds of pure water will produce a concentration of

$$\frac{1 \text{ pound of salt}}{1000 \text{ pounds of water}} = \frac{1000 \times 1 \text{ pound of salt}}{1000 \times 1000 \text{ pounds of water}}$$

$$= \frac{1000 \text{ pounds of salt}}{1{,}000{,}000 \text{ pounds of water}}$$

or 1000 parts per million (ppm) of salt. Seawater contains about 35,000 ppm of dissolved solids. The allowable level of DDT in food is 0.05 ppm or

$$\frac{0.05 \text{ lb of DDT}}{1,000,000 \text{ lb of milk}} \quad \text{or} \quad \frac{0.05 \text{ ounces of DDT}}{1,000,000 \text{ ounces of milk}}$$

$$\text{or} \quad \frac{0.05 \text{ grams of DDT}}{1,000,000 \text{ grams of milk}} \text{ (and so on)}$$

The actual amount of DDT in a cup or a quart or even a gallon of milk is measured in millionths of an ounce, but it is stored in human fat tissue, and even amounts as tiny as 10 ppm can cause serious disorders.

31

```
        1.24
33)41.000
     33
     8 0  ←——————  These remainders are the same and therefore further
     6 6              division will produce a repeat of the digits 24 in the
     1 40             quotient.
     1 32
        8
```

$$\frac{41}{33} = 1.242424 \cdots = 1.\overline{24}$$

Converting decimal numbers to fractions is fairly easy.

$$0.4 = \frac{4}{10} \text{ or } \frac{2}{5}$$

$$0.13 = \frac{13}{100}$$

$$0.275 = \frac{275}{1000} = \frac{11}{40} \text{ (reduced to lowest terms)}$$

$$0.035 = \frac{35}{1000} = \frac{7}{200} \text{ (reduced to lowest terms)}$$

Follow this procedure to convert decimal numbers to fractions:

Step 1   **Example**

Write the non-zero digits to the right of the decimal
point as the numerator in the fraction. $0.00325 = ?$

$$\frac{325}{?}$$

Step 2 $0.\underbrace{00325}_{\text{5 digits}} = \frac{325}{\underbrace{100000}_{\text{5 zeros}}}$

In the denominator write 1 followed by as
many zeros as there are decimal digits in the
decimal number.

Step 3

Reduce the fraction to lowest terms. $\dfrac{325}{100000} = \dfrac{13 \times \cancel{25}}{4000 \times \cancel{25}}$

$$= \frac{13}{4000}$$

Write 0.036 as a fraction in lowest terms. Check your work in ▣.

Decimal-Fraction Equivalents

Some fractions are used so often that it is worthwhile to list their
decimal equivalents. Here are those used most often. Both rounded
form and repeating decimal form are given. The lines marked ▲ are
especially useful and should be memorized.

▲$\dfrac{1}{2} = 0.50$

▲$\dfrac{1}{3} = 0.\overline{3}$ or 0.33 rounded $\dfrac{2}{3} = 0.\overline{6}$ or 0.67 rounded

▲$\dfrac{1}{4} = 0.25$ $\dfrac{2}{4} = \dfrac{1}{2} = 0.50$ $\dfrac{3}{4} = 0.75$

▲$\dfrac{1}{5} = 0.20$ $\dfrac{2}{5} = 0.40$ $\dfrac{3}{5} = 0.60$ $\dfrac{4}{5} = 0.80$

$\dfrac{1}{6} = 0.1\overline{6}$ or 0.17 rounded $\dfrac{5}{6} = 0.8\overline{3}$ or 0.83 rounded

$$\frac{1}{8} = 0.125 \qquad \frac{3}{8} = 0.375 \qquad \frac{5}{8} = 0.625 \qquad \frac{7}{8} = 0.875$$

$$\frac{1}{12} = 0.08\overline{3} \qquad \frac{5}{12} = 0.41\overline{6} \qquad \frac{7}{12} = 0.58\overline{3} \qquad \frac{11}{12} = 0.91\overline{6}$$

$$\frac{1}{16} = 0.0625 \qquad \frac{3}{16} = 0.1875 \qquad \frac{5}{16} = 0.3125$$

$$\frac{7}{16} = 0.4375 \qquad \frac{9}{16} = .5625 \qquad \frac{11}{16} = 0.6875$$

$$\frac{13}{16} = 0.8125$$

$$\frac{1}{20} = 0.05 \qquad \frac{1}{25} = 0.04 \qquad \frac{1}{50} = 0.02$$

32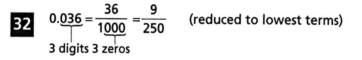

$$0.\underbrace{036}_{\text{3 digits}} = \frac{36}{1\underbrace{000}_{\text{3 zeros}}} = \frac{9}{250} \qquad \text{(reduced to lowest terms)}$$

If the decimal number has a whole number portion, convert the decimal part to a fraction first and then add the whole number part. For example,

$$3.85 = 3 + 0.85$$

$$0.85 = \frac{85}{100} = \frac{17}{20} \qquad \text{(reduced to lowest terms)}$$

$$3.85 = 3 + \frac{17}{20} = 3\frac{17}{20}$$

Write the following decimal numbers as fractions in lowest terms.

(a) 0.0075   (b) 2.08   (c) 3.11

Check your work in **33**.

How to Write a Repeating Decimal as a Fraction

A repeating decimal is one in which some sequence of digits is endlessly repeated. For example, $0.333 \cdots = 0.\overline{3}$ and $0.272727 \cdots = 0.\overline{27}$

are repeating decimals. The bar over the number is a shorthand way of showing that those digits are repeated.

What fraction is equal to $0.\overline{3}$? To answer this, form a fraction with numerator equal to the repeating digits and denominator equal to a number formed with the same number of 9s.

$$0.\overline{3} = \frac{3}{9} = \frac{1}{3}$$

$$0.\overline{27} = \frac{27}{99} = \frac{3}{11}$$ Two digits in $0.\overline{27}$; therefore use 99 as the denominator.

$$0.\overline{123} = \frac{123}{999} = \frac{41}{333}$$ Three digits in $0.\overline{123}$; therefore use 999 as the denominator.

33 (a) $0.0075 = \dfrac{75}{10000} = \dfrac{3}{400}$

(b) $2.08 = 2 + \dfrac{8}{100} = 2 + \dfrac{2}{25} = 2\dfrac{2}{25}$

(c) $3.11 = 3 + \dfrac{11}{100} = 3\dfrac{11}{100}$

Now turn to **34** for a set of practice problems on decimal fractions.

Square Root of a Decimal Number

As you learned in Chapter 1, the **square root** of a number is that number which when squared produces the original number. For example, the square root of 25, written $\sqrt{25}$, is

$$\sqrt{25} = 5 \text{ since } 5 \times 5 = 25$$

The square root of a number is not always a whole number or even a terminating decimal, and so we must approximate it. For example,

$$\sqrt{3} \approx 1.732 \qquad \text{since } 1.732 \times 1.732 = 2.9998 \approx 3$$

and

$$\sqrt{0.07} \approx 0.265 \qquad \text{since } 0.265 \times 0.265 \approx 0.0702 \approx 0.07$$

To find the square root of any number you can use a table of square roots like this:

$2 = 1.4142$
$3 = 1.7321$
$4 = 2$
$5 = 2.2361$
$6 = 2.4495$
$7 = 2.6457$
$8 = 2.8284$
$9 = 3$
$10 = 3.1623$

For most of the numbers shown, the square root continues without end and without repeating. We have rounded the square roots to four decimal places.

Using a calculator is by far the easiest way to find the square root of a number. To calculate a square root, enter the number, then press the $\boxed{\sqrt{}}$ key. For example to calculate $\sqrt{237}$

$237 \boxed{\sqrt{}} \longrightarrow \boxed{15.394804}$

Check: $15.4 \times 15.4 = 237.16$

Try the following problems for practice in using a calculator to find square roots. Round to three decimal places.

1. $\sqrt{7}$	2. $\sqrt{41}$	3. $\sqrt{20}$	4. $\sqrt{10.7}$
5. $\sqrt{1.86}$	6. $\sqrt{70.5}$	7. $\sqrt{110.2}$	8. $\sqrt{51.5}$

Check your answers in the Appendix.

---

34 *Problem Set 3-3:* Decimal Fractions

A. Write as decimal numbers. Round to two decimal digits.

1. $\dfrac{1}{2}$	2. $\dfrac{1}{3}$	3. $\dfrac{2}{3}$
4. $\dfrac{1}{4}$	5. $\dfrac{2}{4}$	6. $\dfrac{3}{4}$
7. $\dfrac{1}{5}$	8. $\dfrac{2}{5}$	9. $\dfrac{3}{5}$
10. $\dfrac{4}{5}$	11. $\dfrac{1}{6}$	12. $\dfrac{5}{6}$

13. $\dfrac{1}{7}$ 14. $\dfrac{2}{7}$ 15. $\dfrac{3}{7}$

16. $\dfrac{1}{8}$ 17. $\dfrac{3}{8}$ 18. $\dfrac{5}{8}$

19. $\dfrac{7}{8}$ 20. $\dfrac{1}{10}$ 21. $\dfrac{2}{10}$

22. $\dfrac{3}{10}$ 23. $\dfrac{1}{12}$ 24. $\dfrac{2}{12}$

25. $\dfrac{3}{12}$ 26. $\dfrac{5}{12}$ 27. $\dfrac{7}{12}$

28. $\dfrac{11}{12}$ 29. $\dfrac{1}{16}$ 30. $\dfrac{3}{16}$

31. $\dfrac{5}{16}$ 32. $\dfrac{7}{16}$ 33. $\dfrac{9}{16}$

34. $\dfrac{11}{16}$ 35. $\dfrac{13}{16}$ 36. $\dfrac{15}{16}$

B. Write as a fraction in lowest terms:

1. 0.3	**2.** 0.75	**3.** 0.44
4. 0.8	**5.** 0.6	**6.** 0.025
7. 0.4	**8.** 1.3	**9.** 2.25
10. 2.05	**11.** 3.16	**12.** 1.125
13. 3.22	**14.** 2.04	**15.** 0.075
16. 10.875	**17.** 0.0007	**18.** 0.0012
19. 0.34	**20.** 11.0105	**21.** 6.0020
22. 4.115	**23.** 0.35	**24.** 0.955

C. Solve:

1. What decimal fraction of 0.5 is 0.6?

2. Find 3.4 of 120.

3. If 0.07 of some number is 0.315, find the number.

4. What decimal part of 2.5 is 42.5?

5. Find a number such that 0.78 of it is 0.390.

D. Brain Boosters

1. (a) $2\frac{3}{5} + 1.785 =$ _____

   (b) $\frac{1}{5} + 1.57 =$ _____

   (c) $3\frac{7}{8} - 2.4 =$ _____

   (d) $1\frac{3}{16} - 0.4194 =$ _____

   (e) $2\frac{1}{2} \times 3.15 =$ _____

   (f) $1\frac{3}{25} \times 2.05 =$ _____

   (g) $3\frac{4}{5} \div 4.75 =$ _____

   (h) $1\frac{5}{16} \div 4.2 =$ _____

2. On four examinations in his history course, Denny scored 73.7, 81.4, 63.6, and 75.6 points. Find his average exam grade. (Add the scores and divide by 4, the number of test scores.)

3. Verify by dividing that $\frac{1}{7} = 0.\overline{142857}$, a repeating decimal. Express $\frac{2}{7}$, $\frac{3}{7}$, $\frac{4}{7}$, $\frac{5}{7}$, and $\frac{6}{7}$ as repeating decimals. What do you notice about these decimal numbers?

4. If seven avocados cost $4, what would be the selling price of one?

5. **Business**   MacDougals Burgers pays its employees $9.72 per hour. What does the company owe Norm for working 23¾ hours?

6. **Business**   The amazing, new, miraculous Microwidgets are priced at 11 for $3.00. What is the price of two amazing, new, miraculous Microwidgets?

7. One tablet of calcium pantothenate contains 0.5 gram. (a) How much is contained in 2¾ tablets? (b) How many tablets are needed to make up 2.6 grams?

E. Calculator Problems
Solve using a calculator.

Write as a decimal number. Round to five decimal digits.

1. $\frac{3}{29}$

2. $\frac{10}{81}$

3. $\frac{11}{17}$

4. $\frac{13}{31}$

5. $\dfrac{23}{41}$ 6. $\dfrac{1357}{2468}$ 7. $\dfrac{19}{23}$ 8. $\dfrac{29}{64}$

Write as a decimal number, rounded to four decimal places.

9. $3\dfrac{17}{19} - 2.15$ 10. $4\dfrac{3}{4} \times 0.0668$

11. $2\dfrac{1}{9} \div 7.126$ 12. $3.205 \div 2\dfrac{1}{17}$

The answers to these problems are in the Appendix. When you have had the practice you need, turn to **35** for a self-test.

35 CHAPTER 3 SELF-TEST

1. $6.2 + 13.045 =$ _____

2. $41.3 + 9.86 =$ _____

3. $16 + 3.407 + 21.744 =$ _____

4. $76 - 7.93 =$ _____

5. $4.27 - 3.8 =$ _____

6. $237.4 - 65.87 =$ _____

7. $90 - 14.85 =$ _____

8. $8.1 \times 2.04 =$ _____

9. $5.6 \times 30 =$ _____

10. $30.4 \times 1.005 =$ _____

11. $8 \div 4.2$ (round to two decimal digits) $=$ _____

12. $14.2 \div 0.075$ (round to two decimal digits) $=$ _____

13. $83.07 \div 104.6$ (round to three decimal digits) $=$ _____

14. Write 0.56 as a fraction in lowest terms. _____

15. Write 3.248 as a fraction in lowest terms. _____

16. Write 32.13 as a fraction in lowest terms. _____

17. Write $\dfrac{7}{16}$ as a decimal. _____

18. Write $3\frac{5}{8}$ as a decimal. _____

19. What part of 3.8 is 4.56? _____

20. What part of 7.0 is 4.2? _____

21. Find 0.25 of 4.8. _____

22. Find 0.65 of 23. _____

23. Find 2.45 of 3.5. _____

24. Find a number such that 0.35 of it is 2.45. _____

25. Find a number such that 1.4 of it is 17.5. _____

The answers to these problems are in the Appendix.

4 Percent

PREVIEW 4

	Where to Go for Help	
When you successfully complete this chapter you will be able to do the following:	Page	**Frame**

1. Write fractions and decimal fractions as percents.

(a) Write 1⅞ as a percent. _____	191	**1**
(b) Write 0.45 as a percent. _____	191	**1**

2. Convert percents to decimal and fraction form.

(a) Write 37½% as a decimal. _____	191	**1**
(b) Write 44% as a fraction. _____	191	**1**

3. Solve problems involving percent.

(a) Find 35% of 16. _____	202	**12**
(b) Find 120% of 45. _____	202	**12**
(c) What percent of 18 is 3? _____	202	**12**
(d) What percent of 1⅓ is ½? _____	202	**12**

	Where to Go for Help	
	Page	Frame

(e) What percent of 0.6 is 0.25? _____   202   **12**

(f) 70% of what number is 56? _____   202   **12**

4. Solve practical problems involving percent.

(a) How much money does a salesperson earn on a $840 sale if his commission is 15%? _____   219   **24**

(b) A camera normally selling for $249.50 is on sale at a discount of 25%. What is its sale price? _____   219   **24**

(c) After a 10% discount a paperback novel sells for $18.45. What was the original price? _____ 219   **24**

(d) A shirt sells for $39.75 plus 6% sales tax. What is the total cost? _____   219   **24**

(e) What is the interest paid on a $1000 bank loan at 9½% annual interest rate for 24 months? _____   219   **24**

If you are certain you can work all of these problems correctly, turn to page 234 for a self-test. If you want help with any of these objectives or if you cannot work one of the preview problems, turn to the page indicated. Super-students who are eager to learn everything in this unit will turn to frame **1** and begin work there.

ANSWERS TO PREVIEW 4 PROBLEMS	**1.** (a) 187.5% (b) 45%	**2.** (a) 0.375 (b) ¹¹⁄₂₅	**3.** (a) 5.6 (b) 54 (c) 16⅔% (d) 30% (e) 41⅔% (f) 80	**4.** (a) $126.00 (b) $187.13 (c) $20.50 (d) $42.14 (e) $190

4 PERCENT

© 1971 United Feature Syndicate, Inc.

1 Numbers and Percent

 If you look through advertisements in the morning newspaper, you may find statements like these:

> "20% off"   "20% discount"   "⅕ savings"
> "Was $50, now on sale for $40"
> "Price slashed 20%"
> "Buy 5 for the price of 4"
> "Buy 4, get 1 free"

All of these "customer lures" are ways of saying the same thing. When buying a car or a house, getting a loan, paying taxes, buying on credit, earning interest on savings, or shopping for a bargain, you need to understand and be able to work with the concept of percent.

Percent The word **percent** comes from the Latin words *per centum* meaning "by the hundred," or "parts per hundred," or "for every hundred." A number expressed as a percent is being compared with a second number called the **Base** standard or **base** by dividing the base into 100 equal parts and writing the comparison number as so many hundredths of the base.

For example, what part of the base or standard length is length A?

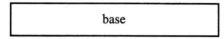

We could answer the question with a fraction or a decimal or a percent. First, divide the base into 100 equal parts. Then compare A with it. The length of A is 40 parts out of the 100 parts that make up the base.

$A = \dfrac{40}{100}$ or 0.40 or 40%.

$\dfrac{40}{100} = 40\%$

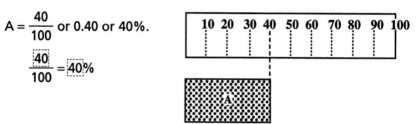

Thus 40% means 40 parts in 100 or $^{40}/_{100}$.

What part of this base is length B? Answer with a percent.

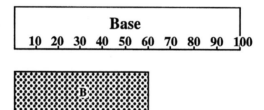

Turn to **2** to check your answer.

2   B is $\dfrac{60}{100}$ or 60%.

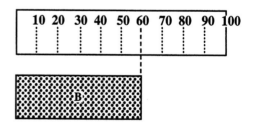

Of course the compared number may be larger than the base.

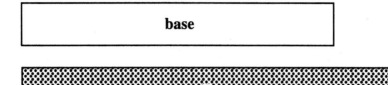

In this case, divide the base into 100 parts and extend it in length.

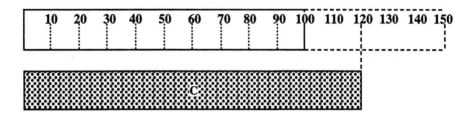

The length of C is 120 parts out of the 100 parts that make up the base.

C is $\dfrac{120}{100}$ or 120% of the base.

Because both our number system and our money system are based on ten and multiples of ten, it is very handy to write comparisons in hundredths or percent.

What part of $1.00 is 50 cents? Write your answer as a fraction, as a decimal, and as a percent. Check in **3** for the answer.

3   *50¢ is what part of $1.00?*

$\dfrac{50¢}{100¢} = \dfrac{50}{100} = 0.50 = 50\%$ 50¢ is equal to 50% of $1.00.

We may also write 50¢ = ½ of $1.00 or 50¢ = 0.50 of $1.00. Fractions, decimals, and percents are alternative ways to talk about a comparison of two numbers.

What percent of 10 is 2? Write 2 as a fraction of 10, rename it as a fraction with denominator equal to 100, then write it as a percent. When you have completed this, go to **4**.

4 $\dfrac{2}{10} = \dfrac{2 \times 10}{10 \times 10} = \dfrac{20}{100} = 20\%$

Decimal to Percent   How do you rewrite a decimal number as a percent? The procedure is simply to multiply the decimal number by 100%.

$$0.60 = 0.60 \times 100\% = 60\%$$

$$
\begin{array}{r}
0.60 \\
\times 100 \\
\hline
60.00 = 60\%
\end{array}
$$

$$0.375 = 0.375 \times 100\% = 37.5\%$$

$$3.4 = 3.4 \times 100\% = 340\%$$

$$0.02 = 0.02 \times 100\% = 2\%$$

Rewrite the following as percents.

(a) 0.70 = _____   (b) 1.25 = _____

(c) 0.001 = _____   (d) 3 = _____

Look in **5** for the answers.

What does "%" mean? Where did that goofy-looking symbol come from?

It means "100." The word "percent" comes from the Latin words meaning "for every hundred." It started as 100, then 1oo, then ⁰⁄₀ and finally % or %.

5   (a) $0.70 = 0.70 \times 100\% = 70\%$

(b) $1.25 = 1.25 \times 100\% = 125\%$

(c) $0.001 = 0.001 \times 100\% = 0.1\%$

(d) $3 = 3 \times 100\% = 300\%$

LEARNING HELP ⬡   Notice in each of these that multiplication by 100% has the effect of moving the decimal point two digits to the right.

0.70 becomes 70.% or 70%

1.25 becomes 125.% or 125%

0.001 becomes 0.1%

3 = 3.00 becomes 300.% or 300%

Fraction to Percent To rewrite a fraction as a percent, we can always rename it as a fraction with 100 as the denominator.

$$\frac{3}{20} = \frac{3 \times 5}{20 \times 5} = \frac{15}{100} = 15\%$$ Multiply by 5 since 100 ÷ 20 = 5

However, the easiest way is to convert the fraction to decimal form by dividing the denominator into the numerator and then multiplying by 100%.

$$\frac{1}{2} = 0.50 = 0.50 \times 100\% = 50\%$$

$$\begin{array}{r} .50 \\ 2\overline{)1.00} \end{array}$$

$$\frac{3}{4} = 0.75 = 0.75 \times 100\% = 75\%$$

$$\begin{array}{r} .75 \\ 4\overline{)3.00} \end{array}$$

$$\frac{3}{20} = 0.15 = 0.15 \times 100\% = 15\%$$

$$\begin{array}{r} .15 \\ 20\overline{)3.00} \\ \underline{2\,0} \\ 1\,00 \\ \underline{1\,00} \end{array}$$

$$1\frac{7}{20} = \frac{27}{20} = 1.35 = 1.35 \times 100\%$$
$$= 135\%$$

$$\begin{array}{r} 1.35 \\ 20\overline{)27.00} \\ \underline{20} \\ 7\,0 \\ \underline{6\,0} \\ 1\,00 \\ \underline{1\,00} \end{array}$$

Rewrite ⁵⁄₁₆ as a percent. Check your answer in **6**.

6 $$\frac{5}{16} = 0.3125 = 0.3125 \times 100\%$$
$$= 31.25\%$$

This is often written as $31\frac{1}{4}\%$.

Remember from Chapter 3 that $0.25 = \frac{1}{4}$.

$$\begin{array}{r} .3125 \\ 16\overline{)5.0000} \\ \underline{4\,8} \\ 20 \\ \underline{16} \\ 40 \\ \underline{32} \\ 80 \\ \underline{80} \end{array}$$

Some fractions cannot be converted to exact decimals. For example,

$$\frac{1}{3} = 0.333 \cdots$$

where the 3s continue to repeat endlessly. We can round to get an approximate percent:

$$\frac{1}{3} \approx 0.333 \times 100\% \approx 33.3\%$$ Approximate percent

Or we can write it as an exact percent:

$$\frac{1}{3} = 0.333 \ldots \times 100\% = 33.333 \ldots \% = 33\frac{1}{3}\% \longleftarrow$$ Exact percent

Rewrite ⅙ as a percent. The answer is in **8**.

7

(a) $\frac{7}{5} = 1.4$ since $5\overline{)7.0}$ (1.4)

Then $\frac{7}{5} = 1.4 \times 100\% = 140\%$

(b) $\frac{2}{3} = 0.666 \ldots$ since $3\overline{)2.000}$ ($0.666 \ldots$)

Then $\frac{2}{3} = 0.666 \ldots \times 100\% = 66.66 \ldots \% = 66\frac{2}{3}\%$ or 66.7% rounded

(c) $3\frac{1}{8} = \frac{25}{8} = 3.125$ since $8\overline{)25.000}$ (3.125)

Then $3\frac{1}{8} = 3.125 \times 100\% = 312.5\%$ or $312\frac{1}{2}\%$

(d) $\frac{5}{12} = 0.41666 \ldots$ since $12\overline{)5.000}$ ($0.416 \ldots$)

Then $\frac{5}{12} = 0.41666 \ldots \times 100\% = 41.666\% = 41\frac{2}{3}\%$

Percent to a Decimal   In order to use percent in solving practical problems, it is often necessary to change a percent to a decimal number. The procedure is to divide by 100%.

For example,

$$50\% = \frac{50\%}{100\%} = \frac{50}{100} = 0.50 \qquad 100\overline{)50.0} \ (.5)$$

$$5\% = \frac{5\%}{100\%} = \frac{5}{100} = 0.05 \qquad 100\overline{)5.00} \;\; {}^{.05}$$

$$0.2\% = \frac{0.2\%}{100\%} = \frac{0.2}{100} = 0.002 \qquad 100\overline{)0.200} \;\; {}^{.002}$$

Fractions may be part of the percent number. If so, it is easiest to reduce to a decimal number and round if necessary.

$$6\frac{1}{2}\% = \frac{6\frac{1}{2}\%}{100\%} = \frac{6.5}{100} = 0.065 \qquad 100\overline{)6.500} \;\; {}^{.065}$$

$$33\frac{1}{3}\% = \frac{33\frac{1}{3}\%}{100\%} = \frac{33\frac{1}{3}}{100} \approx \frac{33.3}{100} = 0.333 \text{ (rounded)}$$

Now you try a few. Write these as decimal numbers.

(a) 4% = _____   (b) 0.5% _____

(c) $16\frac{2}{3}\%$ = _____   (d) $79\frac{1}{4}\%$ = _____

Our answers are in **9**.

---

8 $\frac{1}{6} = 0.1666\ldots$ since $6\overline{)1.0000} \;\; {}^{.1666\ldots}$

$\quad\;\; = 0.1666\ldots \times 100\% = 16.666\ldots\% = 16\frac{2}{3}\%$ exactly

or 16.7% approximately

Rewrite the following fractions as percents.

(a) $\frac{7}{5}$ = _____   (b) $\frac{2}{3}$ = _____

(c) $3\frac{1}{8}$ = _____   (d) $\frac{5}{12}$ = _____

Go to **7** to check your answers.

---

9 (a) $4\% = \frac{4\%}{100\%} = \frac{4}{100} = 0.04 \quad 100\overline{)4.00} \;\; {}^{.04}$

(b) $0.5\% = \dfrac{0.5\%}{100\%} = \dfrac{.5}{100} = 0.005$ $\quad 100\overline{)0.500}^{\,.005}$

(c) $16\dfrac{2}{3}\% = \dfrac{16\dfrac{2}{3}\%}{100\%} = \dfrac{16\dfrac{2}{3}}{100} \approx \dfrac{16.7}{100} = 0.167 \text{ (rounded)}$

(d) $79\dfrac{1}{4}\% = \dfrac{79\dfrac{1}{4}\%}{100\%} = \dfrac{79\dfrac{1}{4}}{100} = \dfrac{79.25}{100} = 0.7925$

Percent to Fraction To change a percent to a fraction, divide by 100% and reduce to lowest terms.

$$36\% = \frac{36\%}{100\%} = \frac{36}{100} = \frac{9 \times 4}{25 \times 4} = \frac{9}{25} \qquad \text{reduced to lowest terms}$$

36% means 36 hundredths or $\dfrac{36}{100}$ or $\dfrac{9}{25}$.

$$12\frac{1}{2}\% = \frac{12\frac{1}{2}\%}{100\%} = \frac{12\frac{1}{2}}{100} = \frac{\frac{25}{2}}{100} = \frac{25}{200} = \frac{1}{8} \qquad \text{reduced to lowest terms}$$

Note that $\dfrac{\frac{25}{2}}{100} = \dfrac{25}{2} \div 100 = \dfrac{25}{2} \times \dfrac{1}{100} = \dfrac{25}{200}.$

$$125\% = \frac{125\%}{100\%} = \frac{125}{100} = \frac{5}{4} = 1\frac{1}{4}$$

For practice, change these percents to fractions.

(a) $72\% = $ _____   (b) $16\dfrac{1}{2}\% = $ _____

(c) $240\% = $ _____   (d) $7\dfrac{1}{2}\% = $ _____ .

You will find the answers in **10**.

It is very helpful to be able to recognize percent–decimal–fraction equivalents. This table should help.

Cheat Sheet

Percent	Decimal	Fraction	Percent	Decimal	Fraction
5%	0.05	$\frac{1}{20}$	50%	0.50	$\frac{1}{2}$
$6\frac{1}{4}\%$	0.0625	$\frac{1}{16}$	60%	0.60	$\frac{3}{5}$
$8\frac{1}{3}\%$	$0.08\overline{3}$	$\frac{1}{12}$	$62\frac{1}{2}\%$	0.625	$\frac{5}{8}$
10%	0.10	$\frac{1}{10}$	$66\frac{2}{3}\%$	$0.6\overline{6}$	$\frac{2}{3}$
$12\frac{1}{2}\%$	0.125	$\frac{1}{8}$	70%	0.70	$\frac{7}{10}$
$16\frac{2}{3}\%$	$0.1\overline{6}$	$\frac{1}{6}$	75%	0.75	$\frac{3}{4}$
20%	0.20	$\frac{1}{5}$	80%	0.80	$\frac{4}{5}$
25%	0.25	$\frac{1}{4}$	$83\frac{1}{3}\%$	$0.8\overline{3}$	$\frac{5}{6}$
30%	0.30	$\frac{3}{10}$	$87\frac{1}{2}\%$	0.875	$\frac{7}{8}$
$33\frac{1}{3}\%$	$0.3\overline{3}$	$\frac{1}{3}$	90%	0.90	$\frac{9}{10}$
$37\frac{1}{2}\%$	0.375	$\frac{3}{8}$	100%	1.00	$\frac{10}{10}$
40%	0.40	$\frac{2}{5}$			

10

(a) $72\% = \dfrac{72\%}{100\%} = \dfrac{72}{100} = \dfrac{18 \times 4}{25 \times 4} = \dfrac{18}{25}$

(b) $16\frac{1}{2}\% = \dfrac{16\frac{1}{2}\%}{100\%} = \dfrac{16\frac{1}{2}}{100} = \dfrac{\frac{33}{2}}{100} = \dfrac{33}{200}$

(c) $240\% = \dfrac{240\%}{100\%} = \dfrac{240}{100} = \dfrac{12 \times 20}{5 \times 20} = \dfrac{12}{5} = 2\dfrac{2}{5}$

(d) $7\dfrac{1}{2}\% = \dfrac{7\dfrac{1}{2}\%}{100\%} = \dfrac{7\dfrac{1}{2}}{100} = \dfrac{\dfrac{15}{2}}{100} = \dfrac{15}{200} = \dfrac{3}{40}$

Now turn to **11** for a set of practice problems on what you have learned in this chapter so far.

11   *Problem Set 4-1:*   Numbers and Percent

A. Write each number as a percent:

1. 0.40	**2.** 0.10	**3.** 0.95	**4.** 0.03
5. 0.3	**6.** 0.015	**7.** 0.6	**8.** 0.01
9. 1.2	**10.** 4.56	**11.** 2.25	**12.** 7.75
13. 0.003	**14.** 3.0	**15.** 0.8	**16.** 5.5
17. 4	**18.** 6.04	**19.** 10	**20.** 0.335

B. Write each fraction as a percent:

1. $\dfrac{1}{5}$	**2.** $\dfrac{3}{4}$	**3.** $\dfrac{7}{10}$	**4.** $\dfrac{7}{20}$
5. $\dfrac{3}{2}$	**6.** $\dfrac{1}{4}$	**7.** $\dfrac{1}{10}$	**8.** $\dfrac{1}{2}$
9. $\dfrac{3}{8}$	**10.** $\dfrac{3}{5}$	**11.** $\dfrac{7}{4}$	**12.** $\dfrac{11}{5}$
13. $1\dfrac{4}{5}$	**14.** $\dfrac{9}{10}$	**15.** $\dfrac{1}{3}$	**16.** $2\dfrac{1}{6}$
17. $\dfrac{2}{3}$	**18.** $\dfrac{11}{16}$	**19.** $\dfrac{23}{12}$	**20.** $3\dfrac{3}{10}$

C. Write each percent as a decimal number:

1. 7%	**2.** 3%	**3.** 56%	**4.** 15%
5. 1%	**6.** $7\dfrac{1}{2}\%$	**7.** 90%	**8.** 200%

9. 0.3%	**10.** 0.07%	**11.** 0.25%	**12.** 150%
13. $1\frac{1}{2}$%	**14.** $6\frac{1}{4}$%	**15.** $\frac{1}{2}$%	**16.** $12\frac{1}{4}$%
17. $125\frac{1}{2}$%	**18.** $66\frac{1}{8}$%	**19.** $30\frac{1}{2}$%	**20.** $8\frac{1}{2}$%

D. Write each percent as a fraction in lowest terms:

1. 10%	**2.** 65%	**3.** 50%	**4.** 20%
5. 25%	**6.** 8%	**7.** 90%	**8.** 135%
9. 3%	**10.** 12%	**11.** $\frac{1}{2}$%	**12.** 0.03%
13. 4.5%	**14.** 220%	**15.** $1\frac{1}{2}$%	**16.** $33\frac{1}{3}$%
17. $7\frac{3}{4}$%	**18.** $6\frac{1}{2}$%	**19.** $16\frac{2}{3}$%	**20.** $3\frac{1}{8}$%

E. Calculator Problems
Write each number as a percent. Round to two decimal places.

1. 1.75623	**2.** 0.05437	**3.** 0.805155
4. 0.14275	**5.** 2.34806	**6.** 12.75261
7. $\frac{7}{13}$	**8.** $\frac{11}{19}$	**9.** $1\frac{7}{17}$
10. $2\frac{3}{7}$	**11.** $\frac{1}{23}$	**12.** $\frac{37}{28}$

The answers to these problems are in the Appendix. When you have had the practice you need, either return to the preview for this chapter on page 189 or continue in frame **12** with the study of problems involving percent.

2   Percent Problems

12   In all of your work with percent you will find that there are three basic types of problems. These three form the basis for all percent problems that arise in business, industry, science, or other areas. All of these problems involve three quantities:

Percentage
- The base or **total (T)** amount or standard used for a comparison
- The **percentage** or **part (P)** being compared with the base or total
- The **percent** or **rate (%)** that indicates the relationship of the percentage to the base (the part to the total)

All three basic percent problems involve finding one of these three quantities when the other two are known. In every problem follow these three steps:

Step 1   **Translate** the problem sentence into a math statement. (If you have not already read the section on word problems in Chapter 2, do so now.) For example, the question "30% of what number is 16?" should be translated:

30% of what number is 16?

$$30\% \times \square = 16$$

In this case, 30% is the percent (%) or rate; $\square$, the unknown quantity, is the total (T) or base; and 16 is the percentage or part (P) of the total. Notice that the words and phrases in the problem become math symbols. The word "of" is translated *multiply*. The word "is" (and similar verbs such as "will be" and "becomes") is translated *equals*. Use a $\square$ or letter of the alphabet or ? for the unknown quantity you are asked to find.

Step 2   It will be helpful if you **label** which numbers are the base or total (T), the percent (%), and the percentage or part (P).

$$30\% \times \square = 16$$
$$\quad\%\qquad T\qquad P$$

Step 3   **Rearrange** the equation so that the unknown quantity is alone on the left of the equal sign and the other quantities are on the right of the equal sign. The equation $30\% \times \square = 16$ becomes

$$\square = 16 \div 30\% = \frac{16}{30\%} = \frac{16}{.30}$$

Step 4   Make a reasonable **estimate** of the answer. *Guess,* but guess carefully. Good guessing is an art.

Step 5   **Solve** the problem by doing the arithmetic.
Never do arithmetic, never multiply or divide, with percent numbers except when converting percents to decimals or decimals to percents. All percents must be rewritten as fractions or decimals before you can use them in a multiplication or division.

Step 6   **Check** your answer against the original guess. Are they the same or at least close? If they do not agree, at least roughly, you have probably made a mistake and should repeat your work.

Step 7   **Double-check** by putting the answer number you have found back into the original problem or equation to see if it makes sense. If possible, use the answer to calculate one of the other numbers in the equation as a check.

Now let's look very carefully at each type of problem. We'll explain each, give examples, show you how to solve them, and work through a few together. Turn to **13**.

13 Type 1   **Type 1 problems** are usually stated in the form "Find 30% of 50" or "What is 30% of 50?" or "30% of 50 is what number?"

Step 1	Translate.	$30\% \times 50 = \square$
Step 2	Label.	$\% \times T = P$
Step 3	Rearrange.	$\square = 30\% \times 50$

You complete the calculation and find $\square$.

$\square = \underline{\hspace{1cm}}$

Choose an answer:

(a) 150   Go to **14**.

(b) 15   Go to **15**.

(c) 1500   Go to **16**.

---

14 You answered that 30% of 50 = 150, and this is not correct. Be certain you do the following in solving this problem:

1. Guess at the answer. A careful, educated guess or estimate is an excellent check on your work. Never work a problem until you know roughly the size of the answer. 30% is roughly ⅓. What is ⅓ of 50, approximately?

2. Never multiply by a percent number. Before you multiply 30% × 50, you must write 30% as a decimal number. (If you need help with this turn to **7**.)

Use these hints to solve the problem. Then return to **13** and choose a better answer.

15 Right you are!

Step 4 *Guess.* The next step is to make an educated guess at the answer. For example,

30% of 50 is roughly $\frac{1}{3}$ of 50 or about 17

Your answer will be closer to 17 than to 2 or 100.

Step 5 *Solve.* Never multiply by a percent number. Percent numbers are not arithmetic numbers. Before you multiply, write the percent number as a decimal number.

$$30\% = \frac{30\%}{100\%} = \frac{30}{100} = 0.30$$

$$30\% \times 50 = 0.30 \times 50 = 15$$

Step 6 *Check.* The guess (17) and the answer (15) are not exactly the same, but they are reasonably close. The answer seems acceptable.

Don't be intimidated by numbers. If the problem involves very large or very complex numbers, reduce it to a simpler problem. The problem

"Find $14\frac{7}{32}\%$ of 6.4"

may look difficult until you realize that it is essentially the same problem as "Find 10% of 6," which is fairly easy.

LEARNING HELP ⬦ Before you begin any actual arithmetic problem involving percent you should:

- Know roughly the size of the answer.
- Have a plan for solving the problem based on a simpler problem.
- Always change percents to decimals or fractions before multiplying or dividing with them.

Now try this problem:

Find $8\frac{1}{2}\%$ of 160.

Check your answer in **17**.

16 Your answer is incorrect.

First, it is important that you make an educated guess at the answer. Never work a problem before you know roughly the size of the answer. In this case, 30% of 50 is roughly ⅓ of 50.

CAUTION ⬦ Second, *never* multiply by a percent number. Before you multiply 30% × 50, you must write 30% as a decimal number. If you need help in writing a percent as a decimal number, turn to frame **7**. Otherwise, return to **13** and try again.

17 Step 1   Translate. $8\frac{1}{2}\% \times 160 = \square$

   Step 2   Label. $\% \times T = P$

Step 3   Rearrange. $\square = 8\frac{1}{2}\% \times 160$

Step 4   Guess.   8½% is close to 10%.

10% of 160 is $\frac{1}{10}$ of 160 or 16.

Step 5   Solve. $8\frac{1}{2}\% = 8.5\% = \frac{8.5}{100} = 0.085$

$\square = 0.085 \times 160 = 13.6$

Step 6   Check.   13.6 is approximately equal to 16, our original estimate.

Now try these for practice.

(a) Find 2% of 140. _____   (b) 35% of 20 = _____

(c) $7\frac{1}{4}\%$ of \$1000 = _____   (d) What is $5\frac{1}{3}\%$ of 3.3? _____

(e) 120% of 15 is what number? _____

The step-by-step answers are in 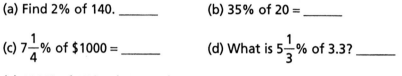.

18   (a) 2% of 140 = ?   Guess: 10% or ⅒ of 140 = 14.

2% × 140 = $\square$ 2% of 140 would be about 3.

$\square = 2\% \times 140$

$2\% = \frac{2\%}{100\%} = \frac{2}{100} = 0.02$

$\square = 0.02 \times 140$
$\square = 2.8$ Check: 2.8 is roughly equal to 3, our original guess.

(b) 35% of 20 = ?

$3\% \times 20 = \Box$

$\Box = 0.35 \times 20$
$\Box = 7$

Guess: 35% is roughly $\frac{1}{3}$, and $\frac{1}{3}$ of 20 is about 6 or 7.

Check: Our guess (6 or 7) is very close to the answer (7).

(c) $7\frac{1}{4}\%$ of $1000 = ?

$7\frac{1}{4}\% \times \$1000 = \Box$

$\Box = 7\frac{1}{4}\% \times \$1000$

$\Box = 0.0725 \times \$1000$

$\Box = \$72.50$

$$7\frac{1}{4}\% = 7.25\% = \frac{7.25}{100} = 0.0725$$

$$\begin{array}{r} 0.0725 \\ \times\ 1000 \\ \hline 72.5000 \end{array}$$

Guess: 10%, or $\frac{1}{10}$, of $1000 is $100. The answer should be less than $100.

Check: The answer, $72.50, is a bit less than the guess of about $100.

(d) $5\frac{1}{3}\%$ of 3.3 = ?

$5\frac{1}{3}\% \times 3.3 = \Box$

$\Box = 5\frac{1}{3}\% \times 3.3$

$\Box = \frac{16}{300} \times 3.3$

$\Box = \frac{16}{300} \times \frac{33}{10} = \frac{528}{3000}$

$\Box = 0.176$

$$5\frac{1}{3}\% = \frac{5\frac{1}{3}}{100} = \frac{\frac{16}{3}}{100} = \frac{16}{300}$$

Divide: $3000\overline{)528.0}$

Guess: 10%, or $\frac{1}{10}$, of 3 is 0.3.

5% of 3 would be half of this or 0.15. A good guess at the answer would be about 0.15.

Check: The answer, 0.176, is reasonably close to the guess: 0.15.

(e) 120% of 15 = ?

$$120\% \times 15 = \square$$
$$\square = 120\% \times 15$$
$$\square = 1.2 \times 15$$
$$\square = 18$$

Guess: 100% of 15 is 15; so the answer is certainly more than 15. 200% of 15 is twice 15 or 30. A good guess would be that the answer is between 15 and 20.

$$120\% = \frac{120}{100} = 1.2$$

Check: The answer, 18, is between 15 and 20, just as we expected our answer to be.

Type 2 Type 2 problems require that you find the rate or percent. Problems of this kind are usually stated "7 is what percent of 16?" or "Find what percent 7 is of 16" or "What percent of 16 is 7?"

Step 1 Translate.

What percent of 16 is 7?
$$\square\% \times 16 = 7$$

Step 2 Label.

$$\% \times T = P$$

All the problem statements are equivalent to this equation.

Step 3 Rearrange.

$$\square\% = \frac{7}{16}$$

To rearrange the equation and solve for $\square\%$, notice that it is in the form

$$\square\% \times 16 = 7$$

Therefore, $\square\% = \dfrac{7}{16}$. Sixteen is the total amount or base, and 7 is the part of the base being described.

Solve this last equation. Check your answer in **19**.

19 **Guess:** $\dfrac{7}{16}$ is very close to $\dfrac{8}{16}$ or $\dfrac{1}{2}$ or 50%. The answer will be a little less than 50%.

$$\square\% = \frac{7}{16} = \frac{7}{16} \times 100\% = \frac{700}{16}\%$$

$$\square\% = 43\frac{3}{4}\%$$

$$\begin{array}{r} 43 \\ 16\overline{)700} \\ 64 \\ \hline 60 \\ 48 \\ \hline 12 \end{array} = 43\frac{12}{16} = 43\frac{3}{4}$$

Check: The answer, $43\frac{3}{4}$%, is reasonably close to our preliminary guess of about 50%.

If you had trouble converting ⁷⁄₁₆ to a percent, you should review this process by turning back to frame **5**.

The solution to a Type 2 problem will be a fraction or decimal number that must be converted to a percent.

Try these problems for practice.

(a) What percent of 40 is 16?

(b) Find what percent 65 is of 25.

(c) $6.50 is what percent of $18.00?

(d) What percent of 2 is 3.5?

(e) $10\frac{2}{5}$ is what percent of 2.6?

Check your work in **20**.

How to Misuse Percent

1. In general you cannot add, subtract, multiply, or divide percent numbers. Percent helps you compare two numbers. It cannot be used in the normal arithmetic operations.

   For example, if 60% of class 1 earned A grades and 50% of class 2 earned A grades, what was the total percent of A grades for the two classes? The answer is that you cannot tell unless you know the number of students in each class.

2. In advertisements designed to trap the unwary, you might hear that "children had 23% fewer cavities when they used . . ." or "50% more doctors use. . . ."

   Fewer than what? Fewer than the worst dental health group the advertiser could find? Fewer than the national average?

   More than what? More than a year ago? More than nurses? More than infants?

   There must be some reference or base given for the percent number to have any meaning at all.

BEWARE of people who misuse percent!

20 (a) $\square\% \times 40 = 16$

$\square\% = \dfrac{16}{40}$

$\dfrac{16}{40} = \dfrac{16}{40} \times 100\%$

$ = \dfrac{1600}{40}\%$

$\square\% = 40\%$

Guess: $\dfrac{16}{40}$ is about $\dfrac{1}{3}$

or roughly 33%.

$\% = \dfrac{P}{T}$

Check: 40% is reasonably close to 33%

Double-check: $40\% \times 40 = ?$
$ 0.40 \times 40 = 16$

(b) $\square\% \times 25 = 65$

$\square\% = \dfrac{65}{25}$

$\dfrac{65}{25} = \dfrac{65}{25} \times 100\%$

$\phantom{\dfrac{65}{25}} = 260\%$

$\square\% = 260\%$

Guess: $\dfrac{65}{25}$ is more than 2 and 2 is 200%. The answer will be over 200%.

Check: The answer agrees with the guess.

Double-check: $260\% \times 25 = ?$
$ 2.60 \times 25 = 65.00 = 65$

The most difficult part of this problem is deciding whether the percent needed is found from ⁶⁵⁄₂₅ or ²⁵⁄₆₅. There is no magic to it. If you read the problem very carefully, you will see that it speaks of 65 as a part "of 25." The base or total is 25. The percentage or part is 65.

(c) $\$6.50 = \square\% \times \$18.00$
or $\square\% \times \$18.00 = \$6.50$

$\square\% = \dfrac{\$6.50}{\$18.00}$

$\% = \dfrac{P}{T}$

Guess: $\$6.50$ is about $\dfrac{1}{3}$ of $\$18.00$

and $\dfrac{1}{3}$ is roughly 33%.

$$\square\% = \frac{6.50}{18.00} = \frac{6.50}{18.00} \times 100\%$$

$$\square\% = \frac{650}{18}\% = 36\frac{2}{18}\%$$

$$\square\% = 36\frac{1}{9}\%$$

Check: $36\frac{1}{9}\%$ is reasonably close to the guess of 33%.

Double-check: $36\frac{1}{9}\% \times \$18 = ?$

$$\frac{325}{900} \times 18 = 6.5$$

(d) $\square\% \times 2 = 3.5$

$$\square\% = \frac{3.5}{2}$$

$$\square\% = \frac{3.5}{2} \times 100\% = \frac{350}{2}\%$$

$$\square\% = 175\%$$

$\% = \dfrac{P}{T}$

Guess: $\dfrac{3}{2}$ is $1\frac{1}{2}$ or 1.5 and 1.5 is 150%. The answer will be something more than 150%.

Check: The answer and the guess are roughly the same.

Double-check: $175\% \times 2 = ?$
$$1.75 \times 2 = 3.50$$

(e) $10\dfrac{2}{5} = \square\% \times 2.6$

or $\square\% \times 2.6 = 10\dfrac{2}{5}$

$$\square\% = \frac{10\frac{2}{5}}{2.6}$$

$$\square\% = \frac{10.4}{2.6} = \frac{10.4}{2.6} \times 100\% = \frac{1040}{2.6}\%$$

$$\square\% = 400\%$$

$\% = \dfrac{P}{T}$

Guess: $\dfrac{10}{2}$ is 5 and 5 is 500%.

Check: The answer and the guess are roughly the same.

Double check: $400\% \times 2.6 = ?$

$$4 \times 2.6 = 10.4 = 10\frac{2}{5}$$

Type 3   Type 3 problems require that you find the total given the percent and the percentage or part. These problems are usually stated like this: "8.7 is 30% of what number?" or "Find a number such that 30% of it is 8.7" or "8.7 is 30% of a number; find the number" or "30% of what number is equal to 8.7?"

Step 1   Translate.   30% of what number is equal to 8.7?
$$30\% \times \square = 8.7$$

Step 2   Label. $\% \times T = P$

Step 3   Rearrange. $\square = \dfrac{8.7}{30\%}$

If $\% \times T = P$
$30\% \times \square = 8.7$

then $\dfrac{P}{\%} = T$

or $T = \dfrac{P}{\%}$

$\square = \dfrac{8.7}{30\%}$

The rearranged problem is $\square = 8.7/30\%$. Solve this problem. Check your answer in **21**.

21 $\square = \dfrac{8.7}{30\%} = \dfrac{8.7}{0.30}$ Guess: $\dfrac{9}{0.3}$ is 30.   ⇨ $0.3\overline{)9.0.0}$ 30.

$\square = 29$ 29.
$0.30\overline{)8.70}$ A reasonable guess is 30.
6 0
2 70   Check: 29 is very close to our guess.
2 70
Double-check: 30% of 29 = ?
$0.30 \times 29 = 8.7$

CAUTION ▷   We cannot divide by 30%. We must change the percent to a decimal number before we do the division.

Here are a few practice problems to test your mental muscles.

(a) 16% of what number is equal to 5.76?

(b) 41 is 5% of what number?

(c) Find a number such that $12\frac{1}{2}\%$ of it is $26\frac{1}{4}$.

(d) 2 is 8% of a number. Find the number.

(e) 125% of what number is 35?

Check your answers against ours in **22**.

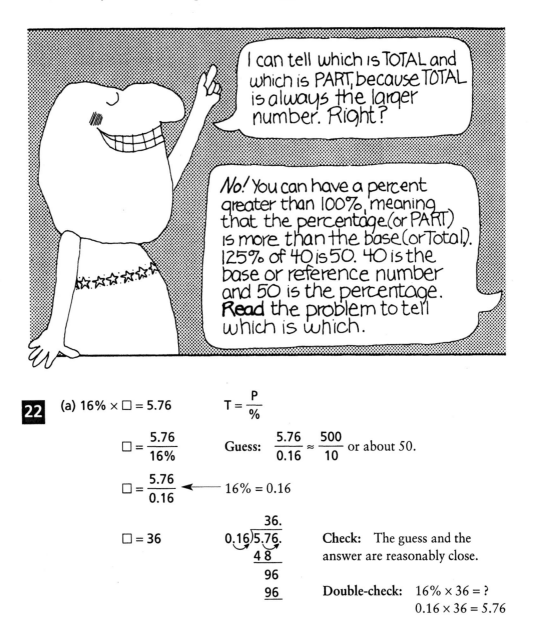

I can tell which is TOTAL and which is PART, because TOTAL is always the larger number. Right?

No! You can have a percent greater than 100%, meaning that the percentage (or PART) is more than the base (or Total). 125% of 40 is 50. 40 is the base or reference number and 50 is the percentage. **Read** the problem to tell which is which.

22 (a) $16\% \times \square = 5.76$

$T = \dfrac{P}{\%}$

$\square = \dfrac{5.76}{16\%}$

Guess: $\dfrac{5.76}{0.16} \approx \dfrac{500}{10}$ or about 50.

$\square = \dfrac{5.76}{0.16}$ ← $16\% = 0.16$

$\square = 36$

$$0.16\overline{)5.76}$$
$$\begin{array}{r} 36. \\ \underline{4\,8} \\ 96 \\ \underline{96} \end{array}$$

Check: The guess and the answer are reasonably close.

Double-check: $16\% \times 36 = ?$
$0.16 \times 36 = 5.76$

(b) $41 = 5\% \times \square$

$$\square = \frac{41}{5\%}$$

$$\square = \frac{41}{0.05}$$

$$\square = 820$$

$$0.05\overline{)41.00.}$$
$$\begin{array}{r} 8\ 20. \\ \underline{40\quad} \\ 1\ 0 \\ \underline{1\ 0} \\ 00 \\ \underline{00} \end{array}$$

$$T = \frac{P}{\%}$$

Guess: $\dfrac{40}{0.05} \approx \dfrac{4000}{5}$ or about 800.

Check: The guess and answer are very close.

Double-check: $5\% \times 820 = ?$
$$0.05 \times 820 = 41$$

(c) $26\dfrac{1}{4} = 12\dfrac{1}{2}\% \times \square$

$$T = \frac{P}{\%}$$

Guess: $\dfrac{26}{0.1}$ is $\dfrac{260}{1}$ or 260.

or $\square = \dfrac{26\dfrac{1}{4}}{12\dfrac{1}{2}\%}$

$$\boxed{12\dfrac{1}{2}\% = 12.5\% = 0.125}$$

$$\square = \frac{26.25}{0.125}$$

Check: 210 and 260 are fairly close.

$$\square = 210$$

$$0.125\overline{)26.250.}$$
$$\begin{array}{r} 210. \\ \underline{25\ 0} \\ 1\ 25 \\ \underline{1\ 25} \\ 00 \\ \underline{00} \end{array}$$

Double-check: $12\dfrac{1}{2}\% \times 210 = ?$

$$0.125 \times 210 = 26.25 = 26\dfrac{1}{4}$$

(d) $2 = 8\% \times \square$

$$T = \frac{P}{\%}$$

or $\square = \dfrac{2}{8\%}$

Guess: $\dfrac{2}{0.08} \approx \dfrac{2}{.1} = 20$

$$\square = \frac{2}{0.08}$$

$$0.08\overline{)2.00.}$$
$$\begin{array}{r} 25. \\ \underline{1\ 6} \\ 40 \\ \underline{40} \end{array}$$

$8\% = 0.08$

$$\square = 25$$

Check: 20 and 25 are close enough.

Double-check: $8\% \times 25 = ?$
$$0.08 \times 25 = 2.00$$

(e) $125\% \times \square = 35$

$$T \times \frac{P}{\%}$$

or $\square = \dfrac{35}{125\%}$

$125\% = 1.25$

$\square = \dfrac{35}{1.25}$

$$1.25\overline{)35.00.}$$
$$\underline{25\ 0}$$
$$10\ 00$$
$$\underline{10\ 00}$$

quotient: $28.$

Guess: $\dfrac{35}{1.25}$ is less than 35.

Guess about 30.

$\square = 28$

Check: The guess (30) and answer (28) agree.

Double-check: $125\% \times 28 = ?$
$$1.25 \times 28 = 35$$

The IS-OF Method for Solving Percent Problems

A shortcut way of solving percent problems involves identifying the two numbers that appear in the problem.

The **IS-number** is always found next to the word *is,* or *equals,* or *is equal to.*

The **OF-number** is always found next to the word *of* in the problem.

- If the problem is obviously a Type 1 problem, as shown on page 199, then it will be clear that the numbers should be multiplied.

- Otherwise, divide the IS-number by the OF-number to find the answer. If the answer should be a percent, rewrite the fraction or decimal answer as a percent.

Because this method is a shortcut, it is especially important that the answer be estimated before any calculations are done. Check the answer after the calculation is completed.

Example
What percent of 16 is 7?

OF-number
IS-number

Guess: 8 is 50% of 16; so 7 is a bit less than 50% of 16.

$$\frac{\text{IS-number}}{\text{OF-number}} = \frac{7}{16} = \frac{700}{16}\% = 43\frac{3}{4}\% \text{ or } 43.75\%$$

Check: $43\dfrac{3}{4}\% \times 16 = 0.4375 \times 16$
$$= 7$$

Example   **Guess:**   12 is roughly one-third
What percent is 12 of 32?   of 32; so the answer is about 33%.

$$\frac{\text{IS-number}}{\text{OF-number}} = \frac{12}{32} = \frac{1200}{32}\% = \frac{75}{2}\% = 37\frac{1}{2}\% \text{ or } 37.5\%$$

Check: $37\frac{1}{2}\% \times 32 = 0.375 \times 32$

$= 12$

For practice, identify the IS-number and the OF-number in each of
the following problems and solve it.

1. 16 is what percent of 20?

2. What percent of 8 is 14?

3. 14% of what number is 98?

4. Find what percent 12 is of 5.

5. 0.04 is what percent of 0.15?

6. 7.2 is 24% of what number?

Check your answers in the Appendix.

A Review

Let's review the seven steps for solving percent problems.

Step 1   *Translate* the problem sentence into a math equation.
Step 2   *Label* the numbers as base or total (T), percentage or part (P),
and percent (%).
Step 3   *Rearrange* the math equation so that the unknown quantity is
alone on the left.
Step 4   *Guess.* Get a reasonable estimate of the answer.
Step 5   *Solve* the problem by doing the arithmetic. *Always* change
percent numbers to decimal numbers first.
Step 6   *Check* your answer by comparing it with the guess in step 4.
Step 7   *Double-check* the answer if you can by putting it back into
the original problem to see if it is correct.

Are you ready for a bit of practice on the three basic kinds of percent
problems? Wind your mind and turn to **23** for a problem set.

23 *Problem Set 4-2:*   Percent Problems

A. Solve:

1. 4 is ____% of 5.

2. 15 is ____% of 75.

3. 6% of $150 = ____.

4. What percent of 25 is 16?

5. 20% of what number is 3?

6. 25% of 428 = ____.

7. 8 is what percent of 8?

8. 120% of 45 is ____.

9. 35% of 60 is ____.

10. 9 is 15% of ____.

11. 8 is ____% of 64.

12. 3% of 5,000 = ____.

13. 100% of what number is 59?

14. 2.5% of what number is 2?

15. What percent of 54 is 36?

16. 60 is ____% of 12.

17. 17 is 17% of ____.

18. 13 is what percent of 25?

19. 74% of what number is 370?

20. $8\frac{1}{2}$% of $250 is ____.

B. Solve:

1. 75 is $33\frac{1}{3}$% of ____.

2. $137\frac{1}{2}$% of 5640 is ____.

3. What percent of 10 is 2.5?

4. 21 is $116\frac{2}{3}$% of ____.

5. 6% of $3.29 is ____.

6. 63 is ____% of 35.

7. 12.5% of what number is 20?

8. $33\frac{1}{3}$% of $8.16 = ____.

9. 9.6 is what percent of 6.4?

10. What percent of 28 is 3.5?

11. What percent of 7.5 is 2?

12. $37\frac{1}{2}$% of 12 is ____.

13. $.75 is ____% of $37.50.

14. 0.516 is what percent of 7.74?

15. $6\frac{1}{4}$% of 280 = ____.

16. $2\frac{1}{4}$ is what percent of 9?

17. 1.28 is ____% of 0.32.

18. 42.7 is 10% of ____.

19. 260% of 8.5 is ____.

20. 4.75% of what number is 76?

C. Brain Boosters

1. If you answer 37 problems correctly on a 42-question test, what percent score do you have?

2. Fifty percent more than what number is 25% less than 32?

3. What is 40% of 90% of 140?

4. What is the difference between ½ of 50% of 17.4 and 50% of ½ of 17.4?

5. The population of Boomville increased from 48,200 to 63,850 in two years. What was the percent increase?

6. A marathon runner, weighing a normal 146 pounds, weighed 138 pounds after his 26-mile race. What percent of his body weight did he lose?

7. **Business**   Profits from Foiled.com, Ed's websales site, increased by $24,450 over last year. This is 30% of last year's total sales. What was last year's total?

8. **Real Estate**   Rudolfo bought a house for $184,500 and made a down payment of 20%. What was the actual amount of the down payment?

9. **Finance**   If 5.45% of your salary should be withheld for Social Security, what amount should be withheld from monthly earnings of $945?

10. **Manufacturing**   One dozen megabolts from Zeus's machine shop sell for $32.50 plus 6% sales tax. Calculate (a) the sales tax and (b) the total cost.

D. Calculator Problems
Solve using a calculator. Round each answer to two decimal places.

1. Find 14.381% of 126.5.

2. Find 8.255% of 6.4.

3. 10.85% of what number is 16.5?

4. What percent of 23.766 is 16.8?

5. 254 is what percent of 175.7?

6. What percent of 125,000 is 37,825?

The answers to these problems are in the Appendix. When you have had the practice you need, either return to the preview on page 189 or continue in frame **24** with the study of some applications of percent to practical problems.

---

3 Applications of Percent

Commission

Commission   The simplest practical use of percent is in the calculation of a part (or percentage) of some total. For example, salespersons are often paid on the basis of their success at selling, and they receive a **commission** or share of the sales receipts. Commission is usually described as a percent of sales income.

> **Commission = rate of commission × total sales**

The rate of commission is given as a percent.

Suppose your job as a door-to-door encyclopedia salesperson pays 12% commission on all sales. How much do you earn from the sale of one $600 set of books?

12% of $600 = _____

Commission = 12% × $600 = 0.12 × $600 = $72	**Guess:** 12% is roughly $\frac{1}{10}$, and $\frac{1}{10}$ of $600 is $60.
	Check: $72 ≈ $60

Try this one yourself:

> *At the Happy Bandit Used Car Company, in addition to salary, each salesperson receives a 6% commission on his or her sales. What would a salesperson earn if he or she sold a 1970 Airedale for $8299.95?*

Check your answer in .

25	**6% of $8299.95 = __?__**	**Guess:** 10% of $8300 is $830. The answer will be about half of this or $415.

Commission = 6% × $8299.95
= 0.06 × $8299.95
= $497.9970
or $498.00 rounded

Check: $498 ≈ $415

The salesperson earns a commission of $498 on the sale. If this same salesperson earns $2550 in commissions in a given week, what was his or her sales total for the week?

Translate the question to a basic percent problem and solve it. Our solution is in **27**.

26 | Commission = percent rate × total sales cost

$2000 = □% × $15,000

Guess: 2 is more than 10% of 15. The answer is somewhere between 10 and 20%.

$$□\% = \frac{\$2,000}{\$15,000} = \frac{2}{15}$$

$$□ = 13\frac{1}{3}\%$$

Check: $13\frac{1}{3}\%$ is reasonably close to the guess.

Ready for some practice? Try these.

(a) A real estate salesperson sells a house for $184,500. Her usual commission is 5%. How much does she earn on the sale?

(b) All salespeople in the Ace Junk Store receive $300 per week plus a 2% commission. If you sold $975 worth of junk in a week, what would be your income?

(c) A salesperson at the Wasteland TV Company sold five DVD television sets last week and earned $128.70 in commissions. If his commission is 6%, what does a DVD TV set cost?

The correct solutions are in **28**.

27 Commission = 6% × total

$2550 = 6% × □   Guess: 10% of what equals $2500? About $25,000. The total sales will be almost double this or $40,000 to $50,000.

$$□ = \frac{\$2550}{6\%} = \frac{\$2550}{0.06}$$

□ = $42,500.00   Check: $42,500 is reasonably close to the guess.

The week's sales total was $42,500.

What rate of commission would a salesperson be receiving if he or she sold a boat for $15,000 and received a commission of $2,000? Look in **26** for the answer.

28 (a) Commission = 5% × $184,500   Guess: 10% of $180,000 is $18,000. She'll earn about $9000.

= 0.05 × $184,500

= $9225   Check: $9225 ≈ $9000

(b) Commission = 2% × $975   Guess: 2% of $100 is $2. Then 2% of $975 is almost 10 times this or almost $20.

= 0.02 × $975

= $19.50   Check: $19.50 ≈ $20

Income = $300 + $19.50 = $319.50

(c) Commission = 6% × total

$128.70 = 6% × □

Guess: 6% is about $\frac{1}{16}$ and

16 × $128 is about
$2000; so a TV set
would cost about $400.

$$□ = \frac{\$128.70}{6\%}$$

$$□ = \frac{\$128.70}{0.06}$$

□ = $2145 total sales

Cost per TV set = $2145 ÷ 5,
             = $429

Check: $429 ≈ $400

Turn to for a look at another application of percent.

29

Discount

Discount
Another important kind of percent application involves the idea of discount. In order to stimulate sales, a merchant may offer to sell some item at less than its normal price. A **discount** is the amount of money by which the normal or list price is reduced to get the sales price or net price.

> **List price – discount = net price or sales price**

List Price
Net Price
The discount is the dollar amount reduction in price. The **list price** is the normal, regular, or original price before the discount is subtracted. The **net price** is the new or sales or discount price. The net price is always less than the list price, of course.

The discount is usually given as a percent of the list price.

> **Discount = rate of discount × list price**

Discount Rate
The **discount rate** is a percent number that enables you to calculate the discount as a part of the list price.

Let's try a problem.

> *The list price of a lamp is $38.50. On a special sale it is offered at 20% off. What is the sales price?*

Can you solve this problem? Try, and then turn to **30** for help.

30 Discount = rate × list price

$$= 20\% \times \$38.50$$

$$= 0.20 \times \$38.50$$
$$= \$7.70$$

Sales price = list price − discount
$$= \$38.50 - \$7.70$$
$$= \$30.80$$

Guess: 20% is $\dfrac{1}{5}$ and $\dfrac{1}{5}$ of $38 is about $8. The sales price will be about $30.

Check: $30.80 ≈ $30

Think of it this way:

Ready for another problem?

After a 25% discount, the sales price of a camera is $243. What was its original or list price?

Check your answer in **31**.

31

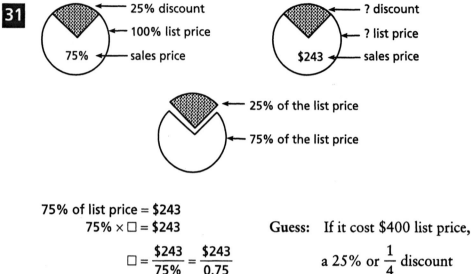

75% of list price = $243

$$75\% \times \square = \$243$$

$$\square = \frac{\$243}{75\%} = \frac{\$243}{0.75}$$

$$\square = \$342$$

Guess: If it cost $400 list price,

a 25% or $\frac{1}{4}$ discount

would give a sales price of about $300. The list price must be a bit less than $400.

Check: $342 is reasonable.

Here are a few problems to test your understanding of the idea of discount.

(a) An after-Christmas sale advertises all toys 70% off. What would be the sales price of a model spaceship that cost $19.95 before the sale?

(b) A clothes dryer is on sale for $376 and is advertised as "12% off regular price." What was its regular price?

(c) A set of four 740-15 automobile tires is on sale for 15% off list price. What would be the sales price if the list price is $87.80 each?

(d) A home computer with a list price of $1200 is offered for sale with a $250 rebate. To what discount rate is this equivalent?

Work hard at these. Knowing how to do them may save you a lot of money one day. Check your answers in **32**.

32

(a) Discount = 70% of $19.95
= 0.70 × $19.95
= $13.965 or
$13.97 (rounded)

Guess: If it is 70% off, it costs 30% of list or about one-third. One-third of $20 is about $6.

Sales price = list price − discount
     = $19.95 − 13.97
     = $5.98   Check: $5.98 ≈ $6

(b) Discount = 12% of list price

Sales price = 88% of list price   Guess: If it cost $400 originally,
     $376 = 88% × □   its sales price would be
                                                   about $300. Its original
                                                   cost was about $400.

$$\square = \frac{\$376}{88\%} = \frac{\$376}{0.88}$$

□ = $427.27 (rounded)   Check: $400 ≈ $427.27

(c) Discount = 15% of list price   Guess: Four tires cost about
     = 15% × (4 × $87.80)   4 × $90, or $360. 10% of
     = 0.15 × $351.20 $360 is $36; so 15% is
     = $52.68   about $54 off. They
                                                   should cost about $300.

Sales price = list price − discount
     = $351.20 − $52.68
     = $298.52   Check: $298.52 ≈ $300

(d) Discount = $250
    Since discount = discount rate × list price, we can write

$$\text{Discount rate} = \frac{\text{discount}}{\text{list price}}$$ Guess: One-fourth of $1200 is
                                                   $300; so $250 is a bit less
                                                   than 25%.

$$= \frac{\$250}{\$1200}$$

≈ 0.208 or   Check: 20.8% ≈ 25%
about 20.8%

Taxes

Taxes are almost always calculated at a percent of some total amount. **Property taxes** are written as some fraction of the value of the property involved. **Income taxes** are most often calculated from complex formulas that depend on many factors. We cannot consider either income or property taxes here.

Sales Tax   A sales tax is an amount calculated from the actual price of a purchase and added to the buyer's cost.

> **Sales tax = tax rate × net price**

Retail sales tax rates are set by the individual states in the United States and vary from 0 to about 9% of the sales price. A sales tax of 6% is often stated as "6¢ on the dollar" since 6% of $1.00 equals 6¢.

Here is an example of a typical tax problem:

If the retail sales tax rate is 7.5% in California, how much would you pay in Los Angeles for a necktie costing $26.50?

Try it; then check your work in **33**.

33

Sales tax = 7.5% of $26.50
       = 0.075 × $26.50
       = $1.99

Guess: 7¢ on each dollar on about $25; then the sales tax should be about 7 × 25 or $1.75. Cost ≈ $28

Actual cost = list price + sales tax
       = $26.50 + $1.99
       = $28.49

Check: $1.99 ≈ $1.75.

Most stores and salesclerks use tables to look up the sales tax and therefore do not need to do the arithmetic shown above except on large purchases beyond the range of the tables. However, it is in your best interest to be able to check their work.

Here are a few problems to test your ability to calculate sales tax. If the sales tax is 5%, find the sales tax on each of the following:

(a) A pen priced at 49¢ _____

(b) A chair priced at $27.50 _____

(c) A toy priced at $2.95 _____

(d) A new car priced at $17,785 _____

(e) A bicycle priced at $226.50 _____

(f) A tube of toothpaste priced at $2.89 _____

Check your answers in **34**.

34	Tax	Total cost	Calculation
	(a) 2¢	51¢	$0.05 \times 49¢ = 2.45¢ \approx 2¢$
	(b) $1.38	$28.88	$0.05 \times 27.50 = \$1.375 \approx \$1.38$
	(c) 15¢	$3.10	$0.05 \times \$2.95 = \$0.1475 \approx 15¢$
	(d) $889.25	$18,674.25	$0.05 \times \$17,785 = \$889.25$
	(e) $11.33	$237.83	$0.05 \times \$226.50 = \$11.325 \approx \$11.33$
	(f) 14¢	$3.03	$0.05 \times \$2.89 = 0.1445 = 14¢$

Interest

In our modern society we have set up complex procedures that allow you to use someone else's money. A *lender* with money beyond his needs supplies cash to a *borrower* whose needs exceed his money. The money is Interest   called a loan. Interest is the amount the lender is paid for the use of his money.

Interest is usually calculated this way:

$$\boxed{\textbf{Interest} = \textbf{principal} \times \textbf{rate of interest} \times \textbf{time}}$$

Principal   The money borrowed is called the *principal,* and the rate of interest is a percent. Unless otherwise stated, the rate of interest is calculated per year and the time is given in years or as a fraction of a year.

The sum of the principal loaned and the interest is called the *amount.*

$$\boxed{\textbf{Amount} = \textbf{principal} + \textbf{interest}}$$

Interest is the money you pay to use someone else's money. The more you use and the longer you use it, the more interest you must pay. The principal is the money you receive, and the amount is the money you must pay back.

When you purchase a house with a bank loan, a car or a refrigerator on an installment loan, or gasoline on a credit card, you are using someone else's money and you pay interest for that use. If you are on the other end of the money game, you may earn interest for money you invest in a savings account or in shares of a business.

Suppose you borrow $500 at 12% annual interest from your local credit union for 1 year. How much interest must you pay? Compute the interest in the following way:

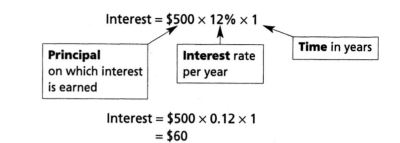

$$\text{Interest} = \$500 \times 0.12 \times 1$$
$$= \$60$$

The amount you repay is

$$\text{Amount} = \$500 + \$60 = \$560$$

Suppose you find yourself in need of cash and arrange to obtain a loan from a bank. You borrow $2600 at 14% per year for 3 months. How much interest must you pay?

Try to set up and solve this problem exactly as we did in the problem above. Our worked solution is in **35**.

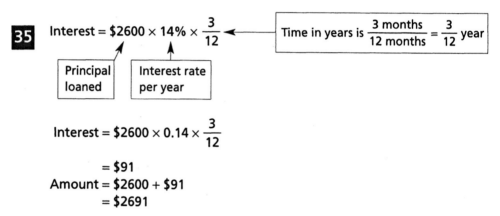

$$\text{Interest} = \$2600 \times 0.14 \times \frac{3}{12}$$

$$= \$91$$
$$\text{Amount} = \$2600 + \$91$$
$$= \$2691$$

Depending on how you and the bank decide to arrange it, you may be required to pay the total principal ($2600) plus interest ($91) all at once at the end of 3 months, or you may use some sort of regular payment plan (for example, pay $897 each month for 3 months).

If you play the money game from the other side of the counter, interest is the amount of money earned by savings or other investment.

For example, if you invest $500 in a savings account in a bank for 6 months at an annual interest rate of 5¼%, you will earn the following interest.

$$\text{Interest} = \$500 \times 5\frac{1}{4}\% \times \frac{6}{12}$$

$$= \$500 \times 0.0525 \times 0.5$$

$$= \$13.1250$$
$$\approx \$13.13 \text{ (rounded)}$$

At the end of 6 months your $500 savings has grown to $513.13.

Your turn. What interest would you earn if you buy a $1200 30-month savings certificate earning 6.34%?

The answer is in **36**.

---

36 $\text{Interest} = \$1200 \times 6.34\% \times \dfrac{30}{12}$

$$= \$1200 \times 0.0634 \times 2.5$$
$$= \$190.20$$

After 30 months your savings certificate is worth

$$\$1200 + \$190.20 = \$1390.20$$

Consumer Credit

Purchasing furniture, appliances, clothing, and similar items from retail stores through an installment plan or deferred payment plan is an American way of life. In this kind of scheme the buyer may make a down payment, receive the items now, and pay the balance later.

The buyer is using the seller's money to make the purchase and is usually charged interest.

Here are three common comsumer credit plans.

- A **charge account** allows the buyer to purchase an item without making a cash outlay. At the end of an agreed period, usually 25 to 30 days, the seller bills the buyer for the cost of the item. No interest charge is made for this service, although there may be a penalty charge for late payment.

- In a **revolving charge account** or **open credit plan,** the buyer makes one or more purchases, and interest is calculated each month on the unpaid balance of the account. It is called a "revolving" credit account because you may continue to make purchases before having repaid the previous balance.

  A **credit card** is a revolving charge account.

- An **installment plan** or multiple payment plan is an agreement to purchase an item and to pay for it with a small down payment and a series of payments spread over a set period of time. It is often used for the purchase of furniture, appliances, or automobiles, where the cost of the item is high and most buyers need to either accumulate the needed cash or spread payments over a period of time.

Here is an example of an installment purchase. You want to buy a new car for $19,600. The dealer agrees to give you $600 for your old clunker as a trade-in and you pay $1000 as a down payment. You borrow the remaining $18,000 from an auto finance company at an interest rate of 12¼%, and it is to be repaid in 36 monthly payments.

Principal = $19,600 − $600 − $1000 = $18,000

Interest = $18,000 × $12\frac{1}{4}$% × 36 months

$$= \$18,000 \times 0.1225 \times 3 \qquad (36 \text{ months} = 3 \text{ years})$$
$$= \$6615$$

You pay $6615 to use the "easy payment plan."

Amount of the loan = $18,000 + $6615
$$= \$24,615 \text{ (total to be repaid)}$$
Monthly payments = $24,615 ÷ 36
$$\approx \$683.75$$

Solve the following problem:

> *A washing machine is on sale for $289.88 and is available on a handy nothing-down, easy payment plan with 24 months to pay. The annual interest rate is 8¼%. What monthly payments would be required?*

Work it out as we did in the example just completed, and then turn to **37** to check your work.

Credit Card Purchases

Using a credit card is equivalent to purchasing on a revolving credit plan. In addition to any annual fees for use of the card, you pay interest at a fixed rate on the unpaid balance of your account.

Suppose you purchase a camera for $200 using a credit card, and the credit card company requires that you repay the loan at a minimum of $25 per month plus interest of 3% per month on the unpaid balance. What interest do you actually pay?

Amount owed	Interest	Payment
Month 1 $200 × 0.03 = $6.00	you pay $25 + $6.00 = $31.00	
Month 2 $175 × 0.03 = $5.25	you pay $25 + $5.25 = $30.25	

Month 3 $150 × 0.03 = $4.50   you pay $25 + $4.50 = $29.50
Month 4 $125 × 0.03 = $3.75   you pay $25 + $3.75 = $28.75
Month 5 $100 × 0.03 = $3.00   you pay $25 + $3.00 = $28.00
Month 6 $ 75 × 0.03 = $2.25   you pay $25 + $2.25 = $27.25
Month 7 $ 50 × 0.03 = $1.50   you pay $25 + $1.50 = $26.50
Month 8 $ 25 × 0.03 = $ .75   you pay $25 + $0.75 = $25.75
$\qquad\qquad\qquad$ $27.00 $\qquad\qquad\qquad\qquad$ $227.00

| Total interest | Total of |
| is $27.00 | 8 payments |

The 3% per month interest rate seems small, but it amounts to about 20% per year.

A bank loan for the $200 at 6% for 8 months would cost you:

$$\$200 \times 6\% \times \frac{8}{12}$$

or

$$\$200 \times 0.06 \times \frac{8}{12}$$

or $8 (quite a difference compared with $27.)

The credit card company demands that you pay more, and in return they are less worried about your ability to meet the payments. For a bigger risk, they want a higher interest.

37

Interest = $289.88 × 8¼% × 2 years
$\qquad$ = $289.88 × 0.0825 × 2
$\qquad$ ≈ $47.83 (rounded)
Amount = $289.88 + $47.83
$\qquad$ = $337.71
Monthly payments = $337.71 ÷ 24
$\qquad$ ≈ $14.07 (rounded)

Percent Increase or Decrease

A very handy way to talk about the change in size of some quantity is to calculate its increase or decrease as a percent of its original size. There are two steps in calculating the percent increase or decrease.

Step 1$\qquad$Find the *amount* of the increase or decrease.
$\qquad\qquad$**Example:** A chain saw cost $250 last year and with inflation it costs $300 now. The amount of increase is

$$\$300 - \$250 = \$50$$

Step 2 Find the percent change as a part of the original amount.

$$\text{Percent increase} = \frac{\text{amount of increase}}{\text{original amount}}$$

or

$$\text{Percent decrease} = \frac{\text{amount of decrease}}{\text{original amount}}$$

Example: For the typewriter, the amount of increase is $50. The original amount is $250.

$$\text{Percent increase} = \frac{50}{250} = 0.20 \text{ or } 20\%$$

Your turn. Sales at the Hall Glass Company dropped from $44,700 last year to $41,960 this year. What percent change is this?
Check your work in **38**.

---

38 Amount of decrease = $44,700 − $41,960 = $2740

$$\text{Percent decrease} = \frac{\text{amount decrease}}{\text{original amount}} = \frac{2740}{44,700}$$

$$= 0.0612 \cdots \text{ or } 6.1\%, \text{ rounded}$$

Now that we have finished our very brief excursion into the mysteries of high finance, turn to **39** for a set of practice problems on these important concepts.

---

39 *Problem Set 4-3:* Practical Applications of Percent

1. A lawyer collected $8500 for a client on a damage suit, and he charged $2480 for his services. What was his rate of commission?

2. **Finance** How much money must you invest at 9% to earn $3000 in a year?

3. **Business** A local sporting goods store offers coaches a 20% discount on all merchandise. The cricket coach at Madam MacAdam's Academy bought a new wicket for $26.96. What was the regular price of the wicket?

4. A laser printer sells for $376 after a 12% discount. What was its original or list price?

5. If you earned $17.50 per hour and received a pay raise to $18.10 per hour, what was the percent pay increase?

6. Net sales for the Apex Cone Company dropped from $76,000 to $71,440 last month. What percent decrease was this?

7. Because of inflation the cost of a T-shirt went from $12 to $14.79. What percent increase is this?

8. **Real Estate**   If you purchased a lot for $25,400 and later sold it for 30% more than its purchase price, what was its selling price?

9. By what percent was a cassette deck reduced if its original price was $389.50 and it was sold for $310.82?

10. **Finance**   To pay for a piece of equipment for his photo studio, Darrell obtained a bank loan for $975 at 9¼% for 9 months. What are his nine monthly payments?

11. If the retail sales tax in your state is 4%, what would be the total cost of each of the following:

(a) a $29,500 sports car
(b) a $1.98 toy
(c) 60¢ worth of nails
(d) a $60.60 textbook
(e) a $5.10 picture frame

12. Students selling tickets to the town carnival received 40¢ for every $4.50 ticket sold. What percent commission is this?

13. A refrigerator has a list price of $729. You buy it while it is on sale at 15% discount and you agree to pay $40 down and the remainder in 12 equal monthly payments at 12% interest per year.

(a) What is the sale price of the refrigerator?
(b) What total interest will you pay?
(c) How big will your payments be?

14. What is the selling price of a set of golf clubs with a list price of $465 if they are on sale at a 35% discount?

15. Sam wants to buy a new car costing $18,000. Being very experienced in money matters he visits several banks, shopping for the best loan. At the First National Bank they offer to finance $17,000 at only 10.5% interest over 3 years. At the Last National Bank they offer to finance only 80% of his car and they want 12.2% interest over 3 years. Which is the better loan? Calculate the interest for each.

16. In a newspaper advertisement a bicycle is offered for sale: "Save $24.70, buy now at 12% off the regular price." What was the regular price?

17. Senator Phil Buster is running for reelection and a recent poll showed that he had 4% of the registered voters in his home county. This week the poll indicates that he has 2% of the voters. The evening TV news said his support is down 2%. Calculate the actual percent decrease in voters.

18. The Double-Talk Bank increased its loan rate from 5% to 6% and advertised this as a 1% increase in rates. Calculate the actual percent increase.

The answers to these problems are in the Appendix. When you have had the practice you need, turn to **40** for a self-test on percent problems.

40 CHAPTER 4 SELF-TEST

1. Write $3\frac{1}{6}$ as a percent. _____

2. Write $\frac{5}{12}$ as a percent. _____

3. Change 0.08 to a percent. _____

4. Write 6.43 as a percent. _____

5. Write 2% as a decimal. _____

6. Write $112\frac{1}{2}$% as a decimal. _____

7. Write 68% as a fraction. _____

8. Find 48% of 250. _____

9. Find 63% of 12. _____

10. Find 165% of 70. _____

11. Find $6\frac{1}{2}$% of 134. _____

12. Find 46.5% of 13.4. _____

13. What percent of 26 is 9.1? _____

14. What percent of 75 is 112.5? _____

15. What percent of 0.72 is 1.62? _____

16. What percent of 1.45 is 0.609? _____

17. What percent of $\frac{3}{8}$ is $\frac{5}{16}$? _____

18. 15% of what number is 12? _____

19. If 120% of a number is 0.84, find the number. _____

20. A gallon of gas costs $1.839. If the price is increased 20%, find the new cost. _____

21. After a 30% discount, an article costs $43.40. Find the original price. __

22. What commission does a salesperson earn on a $3700 sale if her commission rate is 16%? _____

23. What is the sale price of a textbook marked 40% off if its list price is $68.95? _____

24. What is the interest on a $4000 loan at an annual interest of $10\frac{1}{4}\%$ over 30 months? _____

25. If the retail sales tax is 6%, what would be the total cost of a $5.85 toy? _____

The answers to these problems are in the Appendix.

Final Exam

1. $65 + 284 =$ _____

2. $7852 + 519 + 604 =$ _____

3. $403 - 186 =$ _____

4. $7201 - 5325 =$ _____

5. $58 \times 27 =$ _____

6. $354 \times 806 =$ _____

7. $1128 \div 24 =$ _____

8. $38\overline{)17366} =$ _____

9. Factor: $378 =$ _____

10. Factor: $1540 =$ _____

11. $\sqrt{256} =$ _____

12. Write $8\frac{2}{9}$ as an improper fraction: _____

13. Write $\frac{47}{6}$ as a mixed number: _____

14. Reduce to lowest terms: $\frac{660}{924} =$ _____

15. $\dfrac{3}{4} + \dfrac{2}{5} =$ _____

16. $1\dfrac{7}{8} + 2\dfrac{3}{5} =$ _____

17. $2\dfrac{1}{4} + 4\dfrac{5}{6} + 2\dfrac{2}{3} =$ _____

18. $\dfrac{3}{5} - \dfrac{1}{3} =$ _____

19. $4\dfrac{1}{2} - 2\dfrac{3}{7} =$ _____

20. $\dfrac{5}{6} \times \dfrac{3}{4} =$ _____

21. $2\dfrac{4}{7} \times \dfrac{5}{6} =$ _____

22. $\dfrac{3}{8} \div \dfrac{9}{10} =$ _____

23. $3\dfrac{1}{5} \div 2\dfrac{4}{7} =$ _____

24. $\left(4\dfrac{2}{3}\right)^2 =$ _____

25. What fraction of $5\dfrac{2}{3}$ is $1\dfrac{1}{4}$? _____

26. $11.036 + 7.8 =$ _____

27. $129.4 + 6.77 + 28.025 =$ _____

28. $32 - 6.43 =$ _____

29. $8 - 0.27 =$ _____

30. $14.21 - 1.962 =$ _____

31. $9.4 \times 4.3 =$ _____

32. $0.705 \times 48.4 =$ _____

33. $6 \div 9.4$ (round to two decimal digits) = _____

34. $0.0564 \div 0.03 =$ ____

35. Write 4.276 as a fraction in lowest terms: ____

36. Write $\dfrac{5}{6}$ as a decimal: ____

37. What part of 1.7 is 1.36? ____

38. Write $2\dfrac{3}{8}$ as a percent: ____

39. Write 0.01 as a percent: ____

40. Write 128% as a decimal: ____

41. Find 6% of 450. ____

42. Find 125% of 5.6. ____

43. What percent of 56 is 7? ____

44. 32% of what number is 1.6? ____

45. What percent of $\dfrac{2}{3}$ is $\dfrac{1}{5}$? ____

46. At the start of a vacation trip the car odometer reads 23,463 miles. Two weeks later at the end of the trip it reads 27,205 miles. How many miles were traveled on the trip?

47. If sound travels at the rate of 1088 feet per second, how far does it travel in one-half of an hour?

48. If $8510 were divided among 23 people, how much money would each person receive?

49. A $3\dfrac{3}{4}$ pound package of soy nuts costs $5.75. What should 2 pounds cost?

50. Sue carefully counted out the coins in her purse. She had 7 quarters, 12 dimes, 15 nickels, and 29 pennies. How much money did she have in coins?

51. A camera sells for $327 after a 25% discount. What was its original price?

52. A textbook normally sells for $51.45 and is on sale at a 20% discount. What will it cost, including 6% sales tax?

53. What is the cost of 9.7 gallons of gasoline at $1.979 per gallon?

54. What fraction of 60 is $7\frac{1}{2}$?

55. How much money must you have deposited in savings at $7\frac{1}{4}$% in order to earn $137.50 in one year?

56. Gerry's weight dropped from 145 lb to 136 lb while he was on a reducing diet. What percent decrease was this?

Appendix

Boxes and Problem Sets

Chapter 1   Whole Numbers

Box, page 6   Naming Large Numbers

1. Twelve thousand, forty-three

2. Four hundred fifty-seven thousand, nine

3. Two million, twenty-three thousand, sixty-seven

4. One hundred two thousand, four hundred

5. Four million, five thousand, seven

6. Three hundred forty-two million, one hundred three thousand, ten

7. Thirty-four thousand, two

8. Eighty-two million, four thousand, seven hundred

9. Two billion, seven million, fifty-four thousand, two

Box, page 7   Roman Numerals

1. 13	**2.** 17	**3.** 28	**4.** 77	**5.** 124
6. 431	**7.** 1892	**8.** 1910	**9.** 1934	**10.** 1966

Problem Set 1-1, pages 10–11   Practice Problems

A. 8, 16, 6, 13, 11, 11, 13, 9, 13, 12
14, 17, 6, 12, 15, 9, 11, 12, 16, 9
15, 10, 10, 11, 14, 7, 8, 14, 18, 15
14, 14, 12, 12, 16, 12, 13, 13, 13, 15
11, 17, 11, 12, 11, 15, 13, 12, 18, 12

B. 10, 11, 11, 12, 16, 9, 9, 12, 12, 13
10, 13, 15, 8, 14, 9, 13, 18, 15, 11
11, 16, 14, 10, 17, 9, 14, 12, 7, 10
13, 13, 12, 16, 9, 11, 10, 10, 14, 17
11, 14, 11, 10, 11, 14, 16, 12, 13, 15

C. 11, 12, 14, 17, 18, 21, 11, 20, 18, 15
15, 14, 15, 18, 22, 17, 14, 14, 18, 16
12, 19, 8, 10, 22, 12, 19, 15, 14, 16

Box, page 16   How to Add Long Lists of Numbers

1. 91	**2.** 81	**3.** 46	**4.** 92	**5.** 74	**6.** 140

Problem Set 1-2, pages 18–21   Adding Whole Numbers

A. **1.** 70	**2.** 104	**3.** 65	**4.** 126	**5.** 80	**6.** 106
7. 112	**8.** 72	**9.** 131	**10.** 103	**11.** 105	**12.** 123
13. 103	**14.** 124	**15.** 100	**16.** 132	**17.** 52	**18.** 136

B. **1.** 415 **2.** 393 **3.** 1113 **4.** 1003 **5.** 1530

6. 1390 **7.** 1016 **8.** 831 **9.** 1262 **10.** 1009

11. 824 **12.** 806 **13.** 1241 **14.** 861 **15.** 1001

C. **1.** 5525 **2.** 9563 **3.** 9461 **4.** 2611 **5.** 9302

6. 3513 **7.** 3702 **8.** 12,599 **9.** 7365 **10.** 10,122

11. 6505 **12.** 11,428 **13.** 5781 **14.** 15,715 **15.** 9403

16. 11,850

D. **1.** 25,717 **2.** 11,071 **3.** 70,251 **4.** 21,642 **5.** 14,711

6. 89,211 **7.** 47,111 **8.** 175,728 **9.** 101,011 **10.** 180,197

E. **1.** 1042 **2.** 5211 **3.** 2442 **4.** 6441 **5.** 7083

6. 16,275 **7.** 6352 **8.** 7655 **9.** 6514 **10.** 9851

F. **1.** 882 lb **2.** 371 Cal **3.** $2,230,535 **4.** They are the same.

5. 3114 **6.** $70,535 1,083,676,269

G. **1.** 1,377,831 **2.** 104,047 **3.** 551,804 **4.** 41,619

5. 94,666 **6.** 22,428 **7.** 67,275 **8.** 680,212

9. (a) $8200 (b) $7735 (c) Sell

Box, pages 26–27 *Measurement Numbers*

1. 30 ft **2.** 28 lb **3.** 8 ft 3 in. **4.** 3 ft 6 in.

5. 12 lb 1 oz. **6.** 3 lb 2 oz. **7.** 2 hr 5 min **8.** 28 cu ft

9. 26 sq in. **10.** 28 mph

Problem Set 1-3, pages 28–31 *Subtracting Whole Numbers*

A. **1.** 6 **2.** 5 **3.** 7 **4.** 6 **5.** 2 **6.** 5 **7.** 8 **8.** 0

9. 4 **10.** 1 **11.** 9 **12.** 8 **13.** 3 **14.** 7 **15.** 0 **16.** 9

17. 3 **18.** 7 **19.** 3 **20.** 4 **21.** 8 **22.** 1 **23.** 8 **24.** 4

25. 9 **26.** 7 **27.** 9 **28.** 7 **29.** 9 **30.** 4 **31.** 9 **32.** 6

33. 3   34. 1   35. 6   36. 6   37. 8   38. 5   39. 5   40. 8

41. 7   42. 7   43. 18   44. 7   45. 7   46. 3   47. 8   48. 9

B. 1. 13   2. 44   3. 29   4. 16   5. 12   6. 57

7. 19   8. 17   9. 15   10. 28   11. 36   12. 18

13. 22   14. 37   15. 25   16. 26   17. 38   18. 85

C. 1. 189   2. 458   3. 85   4. 877   5. 281   6. 176

7. 154   8. 266   9. 273   10. 198   11. 715   12. 51

13. 574   14. 45   15. 29   16. 145

D. 1. 2809   2. 7781   3. 5698   4. 28,842   5. 12,518

6. 7679   7. 56,042   8. 37,328   9. 4741   10. 9897

11. 9614   12. 26,807   13. 47,593   14. 316,640   15. 22,422

16. 55,459   17. 24,939

E. 1. 1819   2. 284   3. 13,819 mi   4. $23,587   5. $155

6. 2103   7. 3   8. $595

9. $98 - 76 + 54 + 3 + 21$
$123 + 45 - 67 + 8 - 9$
$12 + 3 + 4 + 5 - 6 - 7 + 89$
$123 + 4 - 5 + 67 - 89$
. . . and lots more.

10. Sure, 1,963 pennies are worth $19.63 in fact.

F. 1. 45,805   2. 24,108   3. 588,899   4. 3699

5. 49,799   6. 547,159   7. 313,042   8. 79,508

9. 1,554,108   10. 30,821

Practice Problems, pages 34–35

A. 12, 32, 63, 36, 12, 18, 0, 24, 14, 8
48, 16, 45, 30, 10, 9, 72, 35, 18, 0
28, 15, 36, 49, 8, 40, 42, 54, 64, 24
20, 0, 25, 27, 81, 6, 1, 48, 16, 63

B. 16, 30, 9, 35, 18, 20, 28, 48, 12, 63
32, 0, 18, 24, 9, 25, 24, 45, 10, 72
15, 49, 40, 54, 36, 8, 42, 64, 0, 4
25, 27, 7, 56, 36, 12, 81, 0, 2, 56

Box, pages 42–43   Multiplication by a Multiple of 10/40

1. 110	**2.** 3500	**3.** 4220	**4.** 50,100
5. 65,200	**6.** 723,000	**7.** 403,000	**8.** 8600
9. 214,000	**10.** 2,040,000	**11.** 1,200,300	**12.** 6,000,000
13. 12,080,000	**14.** 720,000	**15.** 105,000,000	

Problem Set 1-4, pages 43–45   Multiplying Whole Numbers

A.

1. 42	**2.** 72	**3.** 56	**4.** 63	**5.** 48	**6.** 54
7. 72	**8.** 42	**9.** 48	**10.** 63	**11.** 56	**12.** 48

B.

1. 87	**2.** 402	**3.** 576	**4.** 243	**5.** 282	**6.** 792
7. 320	**8.** 259	**9.** 156	**10.** 294	**11.** 290	**12.** 261
13. 564	**14.** 392	**15.** 153	**16.** 161	**17.** 282	**18.** 424
19. 308	**20.** 324	**21.** 720	**22.** 1505	**23.** 1728	**24.** 2736
25. 5040	**26.** 2952	**27.** 7138	**28.** 1170	**29.** 1938	**30.** 2548
31. 1650	**32.** 1349	**33.** 4425	**34.** 1458	**35.** 928	**36.** 6232
37. 3822	**38.** 2030	**39.** 8930	**40.** 2752		

C.

1. 37,515	**2.** 74,820	**3.** 375,750	**4.** 97,643
5. 297,591	**6.** 384,030	**7.** 38,023	**8.** 108,486
9. 378,012	**10.** 1,279,840	**11.** 41,064	**12.** 4,947,973
13. 30,780	**14.** 225,852	**15.** 1,368,810	**16.** 31,152
17. 397,584	**18.** 43,381	**19.** 60,241	**20.** 5,098,335
21. 62,006	**22.** 126,000	**23.** 53,212	**24.** 29,500
25. 135,542	**26.** 88,400	**27.** 92,880	**28.** 604,005
29. 361,200	**30.** 1,348,042	**31.** 3,314,204	**32.** 11,033,000
33. 21,135,000	**34.** 3,752,500	**35.** 8,034,030	

D. **1.** $598   **2.** 8760 hr   **3.** $1196   **4.** $1285

    5. 832   **6.** $770,640   **7.** 987,654,321

    8. (a) 111   (b) 1111   (c) 11,111

      (d) 111,111   (e) 1,111,111   (f) 11,111,111

E. **1.** 95,294,763   **2.** 3,501,576,000   **3.** 147,510,720

    4. 23,448,804   **5.** 3,878,280   **6.** 2,423,925

    7. 1,970,640   **8.** 3,864,000,000,000

    9. The products contain only the nine digits 1–9

      (a) 123,458,769   (b) 123,547,689   (c) 967,854,321

10.

49	63
4489	7623
444889	776223
44448889	77762223

Box, page 48   *Averages*

1. 56   **2.** 36   **3.** 10   **4.** 5

5. 21   **6.** 5 hr   **7.** 84 points   **8.** $1664

Box, page 53   *Working with Measurement Numbers*

1. 27 ft   **2.** 110 lb   **3.** 61 ft   **4.** 66 sq ft

5. 25 mph   **6.** 33 m/sec   **7.** 90 mi   **8.** 72 sq ft

9. 180 cu in.   **10.** 186 min

Problem Set 1-5, pages 55–57   *Dividing Whole Numbers*

A. **1.** 9   **2.** 12   **3.** 11 R 4   **4.** 7

    5. Not defined   **6.** 13   **7.** 7 R 2   **8.** 5

    9. 10 R 1   **10.** 7   **11.** 1   **12.** Not defined

    13. 8   **14.** 6   **15.** 4   **16.** 7

    17. 6   **18.** 9

B. 1. 35   2. 41   3. 23 R 6   4. 42   5. 57

   6. 46 R 2   7. 51 R 4   8. 45   9. 112 R 1   10. 44

   11. 21   12. 27   13. 52   14. 37   15. 88

   16. 125   17. 50 R 1   18. 67

C. 1. 23   2. 21   3. 20 R 2   4. 31 R 4   5. 39

   6. 50 R 2   7. 25   8. 19 R 17   9. 9 R 1   10. 41

   11. 53   12. 43   13. 22   14. 11 R 34   15. 12

   16. 34   17. 9 R 6   18. 71 R 5

D. 1. 95 R 6   2. 104   3. 96   4. 208   5. 142 R 6

   6. 107   7. 222 R 2   8. 171   9. 32   10. 1000

   11. 305 R 5   12. 311 R 8   13. 84 R 41   14. 100 R 5   15. 119

   16. 61   17. 102 R 98   18. 81

E. 1. 104 min   2. $1590   3. $560   4. 48 mi/gal

   5. $51   6. 838   7. 248 yr   8. 345 m/sec

F. 1. 4567   2. 1085   3. 302   4. 622

   5. 1001   6. 101   7. 143   8. 124

Box, page 65   Rounding Whole Numbers

1. 50   2. 680   3. 300; 260

4. 7600; 7560   5. 26,400; 26,000   6. 5310; 5300

7. 24,990; 25,000   8. 65,600; 66,000

Problem Set 1-6, pages 70–71   Factors and Factoring

A. 1. $2 \times 2 \times 3$ 2. $2 \times 2 \times 2 \times 2$ 3. 2×7

   4. $2 \times 3 \times 3$ 5. $2 \times 2 \times 2 \times 3$ 6. $2 \times 2 \times 5$

   7. 2×13 8. prime   9. $2 \times 2 \times 2 \times 2 \times 2$

   10. $2 \times 2 \times 3 \times 3$ 11. 3×13 12. $2 \times 3 \times 7$

   13. $2 \times 2 \times 2 \times 7$ 14. $3 \times 3 \times 3 \times 3$ 15. 11×11

B. 1. $2 \times 2 \times 2 \times 2 \times 2 \times 3$ **2.** $2 \times 2 \times 3 \times 7$

3. $2 \times 2 \times 2 \times 17$ **4.** $2 \times 5 \times 17$

5. $2 \times 2 \times 3 \times 3 \times 7$ **6.** $2 \times 2 \times 2 \times 2 \times 2 \times 2 \times 2 \times 2$

7. $2 \times 2 \times 2 \times 2 \times 2 \times 3 \times 3$ **8.** $2 \times 3 \times 5 \times 13$

9. $2 \times 2 \times 3 \times 3 \times 13$ **10.** $2 \times 3 \times 7 \times 13$

11. $2 \times 2 \times 5 \times 7 \times 7$ **12.** 37×37

13. 29×47 **14.** $2 \times 2 \times 3 \times 3 \times 43$

15. 47×67

C. Primes: 2, 5, 3, 31, 23, 37, 53, 19, 67, 61, 89, 17

D. Divisible by 2:   12, 4, 144, 1044, 1390, 72, 102, 2808, 2088, 8280, 8802

Divisible by 3:   9, 12, 231, 45, 144, 261, 1044, 72, 81, 102, 2808, 2088, 8280, 8802, 111

Divisible by 5:   45, 1390, 8280

E. 1. $28 = 1 + 2 + 4 + 7 + 14$

$496 = 1 + 2 + 4 + 8 + 16 + 31 + 62 + 124 + 248$

$8128 = 1 + 2 + 4 + 8 + 16 + 32 + 64 + 127 + 254 + 508 + 1016 + 2032 + 4064$

2.   a. The divisors of 220 sum to 284: $1 + 2 + 4 + 5 + 10 + 11 + 20 + 22 + 44 + 55 + 110 = 284$
The divisors of 284 sum to 220: $1 + 2 + 4 + 71 + 142 = 220$

b. $2924 = 1 + 2 + 4 + 5 + 10 + 20 + 131 + 262 + 524 + 655 + 1310$

$2620 = 1 + 2 + 4 + 17 + 34 + 43 + 68 + 86 + 172 + 731 + 1462$

3.   37; all numbers are primes except for the 1

4.   2, 3, 6

5.   1, not prime; 11, prime; $111 = 3 \times 37$; $1111 = 11 \times 101$;

$11,111 = 41 \times 271$; $111,111 = 3 \times 37 \times 1001$

6.
638	4752	314
475	3658	926
+ 253	4975	+ 705
1366	+ 2403	1945
	15,788	

Problem Set 1-7, pages 79–81   Exponents and Square Roots

A. **1.** 16   **2.** 9   **3.** 64   **4.** 125   **5.** 1000

   6. 49   **7.** 256   **8.** 36   **9.** 512   **10.** 81

   11. 625   **12.** 100,000   **13.** 8   **14.** 243   **15.** 729

   16. 1   **17.** 6   **18.** 1   **19.** 256   **20.** 32

   21. 1,000,000 **22.** 343   **23.** 64   **24.** 1296   **25.** 1024

   26. 6561   **27.** 216   **28.** 25   **29.** 27   **30.** 2401

   31. 1   **32.** 10,000   **33.** 16   **34.** 5   **35.** 4096

   36. 1   **37.** 81   **38.** 1024   **39.** 100   **40.** 7776

B. **1.** 196   **2.** 441   **3.** 3375   **4.** 15,625   **5.** 256

   6. 3025   **7.** 3721   **8.** 64,000   **9.** 1,000,000

   10. 108   **11.** 576   **12.** 1125   **13.** 7938   **14.** 4851

   15. 2025   **16.** 2744   **17.** 1296   **18.** 24,300   **19.** 2000

   20. 90,000   **21.** 9216

C. **1.** 9   **2.** 12   **3.** 4   **4.** 5   **5.** 6   **6.** 10   **7.** 7

   8. 18   **9.** 1   **10.** 11   **11.** 8   **12.** 3   **13.** 15   **14.** 2

   15. 20   **16.** 16

D. **3.** (a) $5^1 = 5$ $5^2 = 25$ $5^3 = 125$ $5^4 = 625$

   (b) $25^1 = 25$ $25^2 = 625$ $25^3 = 15{,}625$ $25^4 = 390{,}625$

   (c) $625^1 = 625$ $625^2 = 390{,}625$ $625^3 = 244{,}140{,}625$
   $625^4 = 152{,}587{,}890{,}625$

   (d) $376^1 = 376$ $376^2 = 141{,}376$ $376^3 = 53{,}157{,}376$
   $376^4 = 19{,}987{,}173{,}376$

Chapter 2   Fractions

Problem Set 2-1, pages 96–98   Renaming Fractions

A. **1.** $\dfrac{7}{3}$ **2.** $\dfrac{22}{5}$ **3.** $\dfrac{15}{2}$ **4.** $\dfrac{94}{7}$ **5.** $\dfrac{35}{4}$ **6.** $\dfrac{4}{1}$

7. $\dfrac{5}{3}$ 8. $\dfrac{35}{6}$ 9. $\dfrac{31}{8}$ 10. $\dfrac{13}{5}$ 11. $\dfrac{161}{10}$ 12. $\dfrac{635}{9}$

13. $\dfrac{481}{40}$ 14. $\dfrac{170}{11}$ 15. $\dfrac{113}{3}$

B. 1. $8\dfrac{1}{2}$ 2. $7\dfrac{2}{3}$ 3. $1\dfrac{3}{5}$ 4. $4\dfrac{3}{4}$ 5. $6\dfrac{1}{6}$ 6. $9\dfrac{1}{3}$

7. $4\dfrac{5}{8}$ 8. $4\dfrac{1}{7}$ 9. $1\dfrac{9}{25}$ 10. $5\dfrac{2}{9}$ 11. $52\dfrac{3}{4}$ 12. $7\dfrac{9}{23}$

13. $4\dfrac{3}{10}$ 14. $20\dfrac{5}{6}$ 15. $9\dfrac{4}{15}$

C. 1. $\dfrac{13}{15}$ 2. $\dfrac{4}{5}$ 3. $\dfrac{4}{5}$ 4. $\dfrac{1}{2}$ 5. $\dfrac{1}{8}$ 6. $\dfrac{2}{5}$

7. $\dfrac{1}{6}$ 8. $\dfrac{8}{9}$ 9. $\dfrac{1}{3}$ 10. $\dfrac{3}{8}$ 11. $\dfrac{7}{20}$ 12. $\dfrac{3}{8}$

13. $\dfrac{1}{6}$ 14. $\dfrac{4}{7}$ 15. $\dfrac{5}{9}$

D. 1. $\dfrac{14}{16}$ 2. $\dfrac{27}{45}$ 3. $\dfrac{9}{12}$ 4. $\dfrac{145}{60}$ 5. $\dfrac{7}{63}$ 6. $\dfrac{45}{35}$

7. $\dfrac{20}{32}$ 8. $\dfrac{140}{25}$ 9. $\dfrac{39}{78}$ 10. $\dfrac{34}{51}$ 11. $\dfrac{363}{44}$ 12. $\dfrac{82}{14}$

13. $\dfrac{66}{72}$ 14. $\dfrac{185}{50}$ 15. $\dfrac{516}{54}$

E. 1. 5 laps, where each lap is $\dfrac{1}{8}$ of a mile.

3. Sugar Glops   4. Too little   5. 100   6. $15\dfrac{3}{4}$ in.   7. $\dfrac{13}{64}$ in.

Problem Set 2-2, pages 102–105   ***Multiplying Fractions***

A. 1. $\dfrac{1}{8}$ 2. $\dfrac{1}{9}$ 3. $\dfrac{4}{15}$ 4. $\dfrac{1}{8}$ 5. $\dfrac{2}{15}$ 6. $\dfrac{5}{6}$

7. 3   **8.** $\frac{1}{2}$ **9.** $2\frac{2}{3}$ **10.** $\frac{5}{6}$ **11.** $\frac{11}{45}$ **12.** $\frac{9}{56}$

13. $1\frac{1}{9}$ **14.** $10\frac{1}{2}$ **15.** $\frac{13}{16}$ **16.** $\frac{4}{5}$ **17.** $2\frac{1}{2}$ **18.** 6

19. 14   **20.** $1\frac{3}{7}$ **21.** $1\frac{1}{3}$

B. 1. 3   **2.** 4   **3.** 8   **4.** $1\frac{5}{16}$ **5.** $3\frac{1}{4}$ **6.** 62

7. $1\frac{1}{21}$ **8.** $\frac{1}{3}$ **9.** 69   **10.** 6   **11.** $35\frac{3}{4}$ **12.** $1\frac{3}{11}$

13. 74   **14.** $7\frac{9}{10}$ **15.** $9\frac{7}{8}$ **16.** $46\frac{2}{3}$ **17.** $10\frac{3}{8}$ **18.** $13\frac{13}{30}$

19. $21\frac{1}{3}$ **20.** 6

C. 1. $\frac{4}{9}$ **2.** $\frac{1}{16}$ **3.** $\frac{27}{125}$ **4.** $10\frac{6}{25}$ **5.** $91\frac{1}{8}$ **6.** $\frac{3}{4}$

7. $\frac{2}{7}$ **8.** $\frac{4}{5}$ **9.** $\frac{5}{8}$ **10.** $\frac{9}{11}$ **11.** $\frac{1}{15}$ **12.** $\frac{1}{6}$

13. $1\frac{1}{3}$ **14.** 20   **15.** $31\frac{1}{2}$

D. 1. 1530 miles   **2.** Bert ate $\frac{1}{2}$ of the pie.

3. 110 km   **4.** $54   **6.** $\frac{7}{8}$ sq mi

7. $16\frac{1}{2}$ mgm   **8.** $128\frac{1}{4}$ in.   **9.** $34\frac{1}{2}$ lb   **10.** $5\frac{2}{5}$ in.   **11.** $393.75

Box, page 112   **Ratio and Proportion**

1. 22   **2.** 462 points   **3.** 198 mi

Problem Set 2-3, pages 112–114   ***Dividing Fractions***

A.   1. $1\frac{2}{3}$ 2. $1\frac{3}{4}$ 3. 9   4. $\frac{1}{12}$ 5. $\frac{5}{16}$

    6. $\frac{4}{9}$ 7. 24   8. $\frac{7}{16}$ 9. $\frac{8}{13}$ 10. $7\frac{1}{2}$

    11. 1   12. $1\frac{1}{2}$ 13. $\frac{5}{28}$ 14. $1\frac{4}{5}$ 15. $2\frac{2}{5}$

B.   1. 9   2. $7\frac{1}{3}$ 3. 4   4. $\frac{3}{4}$ 5. $1\frac{1}{3}$

    6. 6   7. $1\frac{3}{4}$ 8. $2\frac{4}{7}$ 9. $1\frac{1}{5}$ 10. $8\frac{1}{3}$

    11. $1\frac{1}{4}$ 12. $1\frac{2}{3}$ 13. $\frac{6}{7}$ 14. $5\frac{1}{4}$ 15. $\frac{4}{5}$

C.   1. 16   2. $\frac{3}{8}$ 3. $\frac{1}{9}$ 4. $3\frac{1}{9}$ 5. 18

    6. 25   7. $\frac{6}{7}$ 8. 6   9. $1\frac{1}{4}$ 10. $\frac{6}{29}$

    11. $17\frac{1}{2}$ 12. 17

D.   1. $10\frac{8}{13}$ mi   2. $4\frac{1}{2}$ 3. 34 mph   4. $1\frac{3}{8}$ yd   5. 84   6. 400

Problem Set 2-4, pages 127–129   ***Adding and Subtracting Fractions***

A.   1. $1\frac{2}{5}$ 2. $1\frac{1}{3}$ 3. 1   4. $\frac{2}{3}$ 5. $\frac{1}{4}$ 6. $\frac{1}{2}$

    7. $\frac{5}{12}$ 8. $\frac{3}{8}$ 9. $1\frac{1}{8}$ 10. $1\frac{1}{8}$ 11. $\frac{3}{4}$ 12. $\frac{1}{4}$

    13. $1\frac{3}{8}$ 14. $1\frac{2}{3}$ 15. $1\frac{4}{7}$ 16. $\frac{5}{8}$ 17. $1\frac{1}{4}$ 18. $\frac{9}{20}$

B. 1. $1\frac{5}{8}$ 2. $\frac{1}{8}$ 3. $1\frac{11}{36}$ 4. $\frac{19}{36}$ 5. $\frac{29}{48}$ 6. $\frac{29}{35}$

7. $\frac{53}{192}$ 8. $\frac{215}{216}$ 9. $\frac{13}{48}$ 10. $\frac{4}{35}$ 11. $\frac{5}{96}$ 12. $\frac{83}{216}$

13. $1\frac{5}{8}$ 14. $4\frac{1}{36}$ 15. $5\frac{17}{48}$ 16. $4\frac{51}{56}$ 17. $\frac{3}{4}$ 18. $1\frac{23}{48}$

C. 1. $1\frac{2}{3}$ 2. $1\frac{13}{16}$ 3. $5\frac{1}{4}$ 4. $1\frac{4}{5}$ 5. $20\frac{3}{4}$ 6. $1\frac{89}{120}$

7. $\frac{33}{40}$ 8. $1\frac{17}{60}$ 9. $3\frac{1}{12}$ 10. $2\frac{13}{40}$ 11. $5\frac{11}{48}$ 12. $1\frac{9}{10}$

D. 1. $\frac{15}{56}$ in.   2. $36\frac{2}{3}$ hr   4. $\frac{7}{8} = \frac{1}{2} + \frac{1}{4} + \frac{1}{8}$ $\frac{5}{9} = \frac{1}{3} + \frac{1}{6} + \frac{1}{18}$

$$\frac{5}{12} = \frac{1}{3} + \frac{1}{12}$$

5. $414\frac{7}{8}$ ft   6. $28\frac{2}{5}$ gallons; 65 miles per gallon

7. $2\frac{3}{16}$ 8. (b) $\frac{1}{4}$ minus $\frac{1}{4}$ of $\frac{1}{4}$ 9. $23\frac{1}{2}$ hr   10. $88\frac{3}{8}$ acres

Problem Set 2-5, pages 139–141   Solving Word Problems

A. 1. $=$ 2. $\times$ 3. $+$ 4. $=$ 5. $\square - 6$

6. $\frac{1}{2} \times \square$ 7. $2 \times \square$ 8. $\square \div 2\frac{1}{2}$ 9. $\square + \frac{2}{3}$ 10. $\square + \frac{2}{5}$

11. $\square \div \frac{3}{7}$ 12. $\frac{7}{8} \times \square = 1\frac{1}{2}$ 13. $\square - 1\frac{3}{4}$

14. $\square \times 3\frac{1}{4} = 11\frac{1}{2}$ 15. $\frac{7}{16} \times 3\frac{5}{8}$

B. 1. $1\frac{1}{2}$ 2. $\frac{3}{10}$ 3. $1\frac{3}{5}$ 4. $\frac{16}{21}$ 5. $1\frac{3}{8}$

6. $1\frac{1}{4}$ 7. $14\frac{2}{5}$

C. 1. $52   **2.** $8\frac{1}{8}$ miles   **3.** $5\frac{1}{3}$ gallons   **4.** $4\frac{1}{8}$ lb   **5.** $186

Chapter 3   Decimals

Problem Set 3-1, pages 153–156   Adding and Subtracting Decimal Numbers

A. **1.** 0.8   **2.** 1.6   **3.** 1.5   **4.** 0.6   **5.** 0.9   **6.** 1.2

7. 1.6   **8.** 0.7   **9.** 1.7   **10.** 0.1   **11.** 0.7   **12.** 3.3

13. 0.5   **14.** 2.3   **15.** 1.8   **16.** 1.4   **17.** 3.3   **18.** 5.2

19. 1.9   **20.** 9.9   **21.** 4.8   **22.** 1.4   **23.** 9.0   **24.** 18.1

25. 5.5   **26.** 1.4   **27.** 17.5   **28.** 1.6   **29.** 6.7   **30.** 0.6

31. 3.6   **32.** 1.3   **33.** 1.8   **34.** 2.6   **35.** 2.6   **36.** 0.4

B. **1.** 21.01   **2.** $15.02   **3.** 1.617   **4.** 27.19

5. $30.60   **6.** 6.486   **7.** 78.17   **8.** $151.11

9. 5.916   **10.** 828.60   **11.** 16.2019   **12.** 1031.28

13. 63.7313   **14.** 238.24   **15.** 128.3685   **16.** 45.195

17. $27.59   **18.** 70.871   **19.** $108.37   **20.** 19.37

21. $15.36   **22.** 51.34   **23.** 1.04   **24.** $3.86

25. 42.33   **26.** 6.63   **27.** $6.52   **28.** 6.42

29. $36.18   **30.** 22.016   **31.** 2.897   **32.** $17.65

33. 6.96   **34.** 0.3759

C. **1.** 151.461   **2.** 602.654   **3.** 95.888   **4.** 91.15

5. 316.765   **6.** 14.67755   **7.** 16.0425   **8.** 4035.4933

9. 19.011   **10.** 3.34974

D. **1.** $25.14   **2.** 4687.8 mi   **3.** 1112.5 mi   **4.** 2.556 yd

5. 2.267 in.   **6.** $669.35   **7.** 968.749 meters

8. 0.013 cm   **9.** $19,759.97

E. **1.** 25,469.55   **2.** $914.04   **3.** 81,675.83   **4.** 535,258.13

5. $61.23   **6.** 150.666231   **7.** 24,689.082   **8.** 46,807.67

9. 417,109.9   **10.** 121.676   **11.** 278.227   **12.** $372.11

13. $646.30   **14.** 0.163826 sec

Box, pages 168–169   *Multiplying and Dividing by Powers of Ten*

1. 423.8   **2.** 52,923.7   **3.** 622.6   **4.** 177.74

5. 2440.1   **6.** 5003.7   **7.** 23.88   **8.** 48.05

9. 40   **10.** 1800   **11.** 2.466   **12.** 7.0347

13. 2.37882   **14.** 51.702921   **15.** 0.00457   **16.** 0.0002792

17. 0.0002   **18.** 0.37   **19.** 0.000045   **20.** 0.057

Problem Set 3-2, pages 169–173   *Multiplying and Dividing Decimal Numbers*

A. **1.** 0.00001   **2.** 0.1   **3.** 21.5   **4.** 0.06

   5. 0.008   **6.** 0.014   **7.** 0.09   **8.** 0.72

   9. 0.84   **10.** 0.009   **11.** 0.00006   **12.** 4.2

   13. 1.4743   **14.** 2.18225   **15.** 0.5022   **16.** 0.03

   17. 0.024   **18.** 2.16   **19.** 0.01476   **20.** 3.6225

   21. 1.44   **22.** 80.35   **23.** 0.00117   **24.** 287.5

   25. 0.03265   **26.** 0.0009255   **27.** 1223.6   **28.** 0.0001817

B. **1.** 1300   **2.** 12.6   **3.** 0.045   **4.** 13

   5. 126   **6.** 450   **7.** 60   **8.** 2000

   9. 10,000   **10.** 0.037   **11.** 11.1   **12.** 6.6

   13. 1900   **14.** 3256.25   **15.** 0.11   **16.** 2.25

C. **1.** 3.33   **2.** 1.43   **3.** 0.83   **4.** 0.09   **5.** 10.53

   6. 0.67   **7.** 0.25   **8.** 0.13   **9.** 11.11   **10.** 285.71

   11. 0.15   **12.** 16.00   **13.** 14.286   **14.** 0.023   **15.** 0.225

   16. 65   **17.** 3.462   **18.** 13.640   **19.** 3.815   **20.** 2.999

   21. 571.429   **22.** 1111

D. **1.** 0.09   **2.** 0.0009   **3.** 0.000009   **4.** 0.027

5. 1.44   **6.** 1.728   **7.** 0.01   **8.** 0.0001

9. 0.000027   **10.** 0.012   **11.** 0.143   **12.** 0.223

13. 0.019   **14.** 0.215   **15.** 138.889

E. **1.** $233.75   **2.** $23.11   **3.** $126.00   **4.** 26.6   **5.** 0°F

6. 1st: (b) 12 ÷ 0.03 = 400, 2nd: 0.2 × 0.2 = 0.04, 3rd: The third
error is that there are only two errors.

7. January 1   **8.** Yes. It is also true for any other digit.

9. (a) 12.5 oz   (b) 47 oz   (c) $8.28   **10.** 6255 lb

F. **1.** 1.375   **2.** 6.066   **3.** 2.082   **4.** 6.062   **5.** 0.667

6. 0.099   **7.** 6.334   **8.** 4.112   **9.** 1.8182   **10.** 0.0250

11. 0.2857   **12.** 246.1624   **13.** 0.1111   **14.** 0.0390   **15.** 2.1866

16. 27.4703

Box, page 183   *Square Root of a Decimal Number*

1. 2.646   **2.** 6.403   **3.** 4.472   **4.** 3.271   **5.** 1.364

6. 8.396   **7.** 10.498   **8.** 7.176

Problem Set 3-3, pages 183–186   *Decimal Fractions*

A. **1.** 0.50   **2.** 0.33   **3.** 0.67   **4.** 0.25   **5.** 0.50   **6.** 0.75

7. 0.20   **8.** 0.40   **9.** 0.60   **10.** 0.80   **11.** 0.17   **12.** 0.83

13. 0.14   **14.** 0.29   **15.** 0.43   **16.** 0.13   **17.** 0.38   **18.** 0.63

19. 0.88   **20.** 0.10   **21.** 0.20   **22.** 0.30   **23.** 0.08   **24.** 0.17

25. 0.25   **26.** 0.42   **27.** 0.58   **28.** 0.92   **29.** 0.06   **30.** 0.19

31. 0.31   **32.** 0.44   **33.** 0.56   **34.** 0.69   **35.** 0.81   **36.** 0.94

B. **1.** $\dfrac{3}{10}$ **2.** $\dfrac{3}{4}$ **3.** $\dfrac{11}{25}$ **4.** $\dfrac{4}{5}$

5. $\dfrac{3}{5}$ **6.** $\dfrac{1}{40}$ **7.** $\dfrac{2}{5}$ **8.** $1\dfrac{3}{10}$

9. $2\frac{1}{4}$

10. $2\frac{1}{20}$

11. $3\frac{4}{25}$

12. $1\frac{1}{8}$

13. $3\frac{11}{50}$

14. $2\frac{1}{25}$

15. $\frac{3}{40}$

16. $10\frac{7}{8}$

17. $\frac{7}{10000}$

18. $\frac{3}{2500}$

19. $\frac{17}{50}$

20. $11\frac{21}{2000}$

21. $6\frac{1}{500}$

22. $4\frac{23}{200}$

23. $\frac{7}{20}$

24. $\frac{191}{200}$

C. 1. 1.2   **2.** 408   **3.** 4.5   **4.** 17   **5.** 0.5

D. 1. (a) 4.385   (b) 1.77   (c) 1.475   (d) 0.7681

(e) 7.875   (f) 2.296   (g) 0.8   (h) 0.3125

2. 73.575, round to 73.6

3. The same six digits repeat. $\frac{2}{7} = 0.\overline{285714}$, $\frac{3}{7} = 0.\overline{428571}$

$\frac{4}{7} = 0.\overline{571428}$, $\frac{5}{7} = 0.\overline{714285}$, $\frac{6}{7} = 0.\overline{857142}$

4. 57¢   **5.** $230.85   **6.** 55¢   **7.** (a) 1.375 grams, (b) $5\frac{1}{5}$ tablets

E. 1. 0.10345   **2.** 0.12346   **3.** 0.64706   **4.** 0.41935

5. 0.56098   **6.** 0.54984   **7.** 0.82609   **8.** 0.45313

9. 1.7447   **10.** 0.3173   **11.** 0.2963   **12.** 1.5567

Chapter 4   Percent
Problem Set 4-1, pages 200–201   Numbers and Percent

A.   1. 40%   **2.** 10%   **3.** 95%   **4.** 3%   **5.** 30%

6. 1.5%   **7.** 60%   **8.** 1%   **9.** 120%   **10.** 456%

11. 225%   **12.** 775%   **13.** 0.3%   **14.** 300%   **15.** 80%

16. 550%   **17.** 400%   **18.** 604%   **19.** 1000%   **20.** 33.5%

B. **1.** 20%   **2.** 75%   **3.** 70%   **4.** 35%   **5.** 150%

6. 25%   **7.** 10%   **8.** 50%   **9.** $37\frac{1}{2}$%   **10.** 60%

11. 175%   **12.** 220%   **13.** 180%   **14.** 90%   **15.** $33\frac{1}{3}$%

16. $216\frac{2}{3}$%   **17.** $66\frac{2}{3}$%   **18.** $68\frac{3}{4}$%   **19.** $191\frac{2}{3}$%   **20.** 330%

C. **1.** 0.07   **2.** 0.03   **3.** 0.56   **4.** 0.15   **5.** 0.01

6. 0.075   **7.** 0.90   **8.** 2.00   **9.** 0.003   **10.** 0.0007

11. 0.0025   **12.** 1.50   **13.** 0.015   **14.** 0.0625   **15.** 0.005

16. 0.1225   **17.** 1.255   **18.** 0.66125   **19.** 0.305   **20.** 0.085

D. **1.** $\frac{1}{10}$ **2.** $\frac{13}{20}$ **3.** $\frac{1}{2}$ **4.** $\frac{1}{5}$ **5.** $\frac{1}{4}$

6. $\frac{2}{25}$ **7.** $\frac{9}{10}$ **8.** $1\frac{7}{20}$ **9.** $\frac{3}{100}$ **10.** $\frac{3}{25}$

11. $\frac{1}{200}$ **12.** $\frac{3}{10000}$ **13.** $\frac{9}{200}$ **14.** $2\frac{1}{5}$ **15.** $\frac{3}{200}$

16. $\frac{1}{3}$ **17.** $\frac{31}{400}$ **18.** $\frac{13}{200}$ **19.** $\frac{1}{6}$ **20.** $\frac{1}{32}$

E. **1.** 175.62%   **2.** 5.44%   **3.** 80.52%   **4.** 14.28%

5. 234.81%   **6.** 1275.26%   **7.** 53.85%   **8.** 57.89%

9. 141.18%   **10.** 242.86%   **11.** 4.35%   **12.** 132.14%

Box, page 216   IS-OF Method

1. IS = 16, OF = 20, 80%   **2.** IS = 14, OF = 8, 175%

3. IS = 98, OF = 0.14, 700   **4.** IS = 12, OF = 5, 240%

5. IS = 0.04, OF = 0.15, $26\frac{2}{3}$%   **6.** IS = 7.2, OF = 0.24, 30

Problem Set 4-2, pages 217–219   Percent Problems

A. **1.** 80%   **2.** 20%   **3.** $9   **4.** 64%   **5.** 15

   6. 107   **7.** 100%   **8.** 54   **9.** 21   **10.** 60

   11. $12\frac{1}{2}$%   **12.** 150   **13.** 59   **14.** 80   **15.** $66\frac{2}{3}$%

   16. 500%   **17.** 100   **18.** 52%   **19.** 500   **20.** $21.25

B. **1.** 225   **2.** 7755   **3.** 25%   **4.** 18

   5. $0.1974 or 20¢   **6.** 180%   **7.** 160   **8.** $2.72

   9. 150%   **10.** $12\frac{1}{2}$%   **11.** $26\frac{2}{3}$ **12.** 4.5

   13. 2%   **14.** $6\frac{2}{3}$%   **15.** 17.5   **16.** 25%

   17. 400%   **18.** 427   **19.** 22.1   **20.** 1600

C. **1.** 88%   **2.** 16   **3.** 50.4

   4. No difference   **5.** 32.5% rounded   **6.** 5.5% rounded   **7.** $81,500

   8. $36,900   **9.** $51.50   **10.** (a) $1.95   (b) $34.45

D. **1.** 18.19   **2.** 0.53   **3.** 152.07   **4.** 70.69%

   5. 144.56%   **6.** 30.26%

Problem Set 4-3, pages 232–234   Practical Applications of Percent

1. 29%   **2.** $33,333.33   **3.** $33.70   **4.** $427.27

5. 3.4%   **6.** 6%   **7.** 23.25%   **8.** $33,020

9. 20.2%   **10.** $115.85

11. (a) $30,680   (b) $2.06   (c) 62¢   (d) $63.02   (e) $5.30

12. 8.89%   **13.** (a) $619.65   (b) $69.56   (c) $54.10

14. $302.25   **15.** $5355 at First National, $5270.40 at Last National

16. $205.83   **17.** 50%   **18.** 20%

Answers: Chapter Self-Tests

If you answer any question incorrectly, turn to the page and frame indicated for review.

Chapter 1 Self-Test, pages 81–82

	Page	Frame			Page	Frame
1. 83	3	**1**	**14.** 502		46	**35**
2. 5221	3	**1**	**15.** 32		46	**35**
3. 164	3	**1**	**16.** $2^3 \times 3 \times 17$		57	**44**
4. 26	21	**16**	**17.** $2 \times 3 \times 11^2 \times 13$		57	**44**
5. 286	21	**16**	**18.** $2^3 \times 5^3$		57	**44**
6. 2028	21	**16**	**19.** $2^6 \times 3 \times 17$		57	**44**
7. 3695	21	**16**	**20.** 81		72	**59**
8. 1548	32	**24**	**21.** 2000		72	**59**
9. 48,348	32	**24**	**22.** 7938		72	**59**
10. 55,622	32	**24**	**23.** 15,129		72	**59**
11. 650,206	32	**24**	**24.** 15		72	**59**
12. 23	46	**35**	**25.** 18		72	**59**
13. 803 R6	46	**35**				

Chapter 2 Self-Test, pages 141–142

	Page	Frame			Page	Frame
1. $\dfrac{115}{16}$	85	**1**	**5.** $\dfrac{31}{35}$		114	**34**
2. $3\dfrac{4}{11}$	85	**1**	**6.** $1\dfrac{19}{60}$		114	**34**
3. $\dfrac{15}{40}$	85	**1**	**7.** $6\dfrac{7}{24}$		114	**34**
4. $\dfrac{13}{17}$	85	**1**	**8.** $\dfrac{5}{12}$		114	**34**

	Page	Frame		Page	Frame
9. $1\frac{3}{20}$	114	34	18. $\frac{7}{16}$	129	51
10. $3\frac{5}{12}$	114	34	19. $\frac{2}{5}$	129	51
11. 1	98	19	20. $\frac{3}{4}$	129	51
12. $5\frac{1}{7}$	98	19	21. 2	129	51
13. $1\frac{1}{9}$	105	26	22. $\frac{7}{25}$	129	51
14. $\frac{69}{154}$	105	26	23. $12\frac{1}{4}$	129	51
15. $\frac{1}{4}$	98	19	24. $\frac{9}{100}$	129	51
16. $5\frac{4}{9}$	98	19	25. $72	129	51
17. $58\frac{1}{2}$	98	19			

Chapter 3 Self-Test, pages 186–187

	Page	Frame		Page	Frame
1. 19.245	145	1	7. 75.15	145	1
2. 51.16	145	1	8. 16.524	156	10
3. 41.151	145	1	9. 168	156	10
4. 68.07	145	1	10. 30.552	156	10
5. 0.47	145	1	11. 1.90	162	16
6. 171.53	145	1	12. 189.33	162	16

	Page	Frame		Page	Frame
13. 0.794	162	16	20. 0.6	173	25
14. $\frac{14}{25}$	173	25	21. 1.2	173	25
15. $3\frac{31}{125}$	173	25	22. 14.95	173	25
16. $32\frac{13}{100}$	173	25	23. 8.575	173	25
17. 0.4375	173	25	24. 7	173	25
18. 3.625	173	25	25. 12.5	173	25
19. 1.2	173	25			

Chapter 4 Self-Test, pages 234–235

	Page	Frame		Page	Frame
1. $316\frac{2}{3}\%$	191	1	14. 150%	206	18
2. $41\frac{2}{3}\%$	191	1	15. 225%	206	18
3. 8%	193	4	16. 42%	206	18
4. 643%	193	4	17. $83\frac{1}{3}\%$	206	18
5. 0.02	196	7	18. 80	210	20
6. 1.125	196	7	19. 0.7	210	20
7. $\frac{17}{25}$	197	9	20. $2.207	219	24
8. 120	203	13	21. $62.00	219	24
9. 7.56	203	13	22. $592	219	24
10. 115.5	203	13	23. $41.37	222	29
11. 8.71	203	13	24. $1025	227	34
12. 6.231	203	13	25. $6.20	224	32
13. 35%	206	18			

Answers: Final Exam, pages 237–240

1. 349

2. 8975

3. 217

4. 1876

5. 1566

6. 285,324

7. 47

8. 457

9. $2 \times 3^3 \times 7$

10. $2^2 \times 5 \times 7 \times 11$

11. 16

12. $\dfrac{74}{9}$

13. $7\dfrac{5}{6}$

14. $\dfrac{5}{7}$

15. $1\dfrac{3}{20}$

16. $4\dfrac{19}{40}$

17. $9\dfrac{3}{4}$

18. $\dfrac{4}{15}$

19. $\dfrac{29}{14}$ or $2\dfrac{1}{14}$

20. $\dfrac{15}{24}$ or $\dfrac{5}{8}$

21. $2\dfrac{1}{7}$

22. $\dfrac{5}{12}$

23. $1\dfrac{11}{45}$

24. $21\dfrac{7}{9}$

25. $\dfrac{15}{68}$

26. 18.836

27. 164.195

28. 25.57

29. 7.73

30. 12.248

31. 40.42

32. 34.1220

33. 0.64

34. 1.88

35. $4\dfrac{69}{250}$

36. $0.8\overline{3}$

37. 0.8

38. $237\dfrac{1}{2}\%$

39. 1%

40. 1.28

41. 27

42. 7

43. 12.5

44. 5

45. 30%

46. 3742

47. 1,958,400 ft $\cong$ 371 miles

48. $370

49. $3.07

50. $3.99

51. $436

52. $43.63

53. $19.20

54. $\dfrac{1}{8}$

55. $1896.55

56. 6.2%

TABLE OF SQUARES

Number	Square	Number	Square	Number	Square
1	1	35	1225	68	4624
2	4	36	1296	69	4761
3	9	37	1369	70	4900
4	16	38	1444	71	5041
5	25	39	1521	72	5184
6	36	40	1600	73	5329
7	49	41	1681	74	5476
8	64	42	1764	75	5625
9	81	43	1849	76	5776
10	100	44	1936	77	5929
11	121	45	2025	78	6084
12	144	46	2116	79	6241
13	169	47	2209	80	6400
14	196	48	2304	81	6561
15	225	49	2401	82	6724
16	256	50	2500	83	6889
17	289	51	2601	84	7056
18	324	52	2704	85	7225
19	361	53	2809	86	7396
20	400	54	2916	87	7569
21	441	55	3025	88	7744
22	484	56	3136	89	7921
23	529	57	3249	90	8100
24	576	58	3364	91	8281
25	625	59	3481	92	8464
26	676	60	3600	93	8649
27	729	61	3721	94	8836
28	784	62	3844	95	9025
29	841	63	3969	96	9216
30	900	64	4096	97	9409
31	961	65	4225	98	9604
32	1024	66	4356	99	9801
33	1089	67	4489	100	10000
34	1156				

TABLE OF SQUARE ROOTS

Number	Square Root	Number	Square Root	Number	Square Root	Number	Square Root
1	1.0000	51	7.1414	101	10.0499	151	12.2882
2	1.4142	52	7.2111	102	10.0995	152	12.3288
3	1.7321	53	7.2801	103	10.1489	153	12.3693
4	2.0000	54	7.3485	104	10.1980	154	12.4097
5	2.2361	55	7.4162	105	10.2470	155	12.4199
6	2.4495	56	7.4833	106	10.2956	156	12.4900
7	2.6458	57	7.5198	107	10.3441	157	12.5300
8	2.8284	58	7.6158	108	10.3923	158	12.5698
9	3.0000	59	7.6811	109	10.4403	159	12.6095
10	3.1623	60	7.7460	110	10.4881	160	12.6191
11	3.3166	61	7.8102	111	10.5357	161	12.6886
12	3.4640	62	7.8740	112	10.5830	162	12.7279
13	3.6056	63	7.9397	113	10.6301	163	12.7671
14	3.7417	64	8.0000	114	10.6771	164	12.8062
15	3.8730	65	8.0623	115	10.7238	165	12.8452
16	4.0000	66	8.1240	116	10.7703	166	12.8841
17	4.1231	67	8.1854	117	10.8167	167	12.9228
18	4.2426	68	8.2462	118	10.8628	168	12.9615
19	4.3589	69	8.3066	119	10.9087	169	13.0000
20	4.4721	70	8.3666	120	10.9545	170	13.0384
21	4.5826	71	8.4261	121	11.0000	171	13.0767
22	4.6904	72	8.4853	122	11.0454	172	13.1149
23	4.7958	73	8.5440	123	11.0905	173	13.1529
24	4.8990	74	8.6023	124	11.1355	174	13.1909
25	5.0000	75	8.6603	125	11.1803	175	13.2288
26	5.0990	76	8.7178	126	11.2250	176	13.2665
27	5.1962	77	8.7750	127	11.2694	177	13.3041
28	5.2915	78	8.8318	128	11.3137	178	13.3417
29	5.3852	79	8.8882	129	11.3578	179	13.3791
30	5.4772	80	8.9443	130	11.4018	180	13.4164
31	5.5678	81	9.0000	131	11.4455	181	13.4536
32	5.6569	82	9.0554	132	11.4891	182	13.4907
33	5.7446	83	9.1104	133	11.5326	183	13.5277
34	5.8310	84	9.1652	134	11.5758	184	13.5647
35	5.9161	85	9.2195	135	11.6190	185	13.6015
36	6.0000	86	9.2736	136	11.6619	186	13.6382
37	6.0828	87	9.3274	137	11.7047	187	13.6748
38	6.1644	88	9.3808	138	11.7473	188	13.7113
39	6.2450	89	9.4340	139	11.7898	189	13.7477
40	6.3246	90	9.4868	140	11.8322	190	13.7840
41	6.4031	91	9.5391	141	11.8743	191	13.8203
42	6.4807	92	9.5917	142	11.9164	192	13.8564
43	6.5574	93	9.6437	143	11.9583	193	13.8924
44	6.6332	94	9.6954	144	12.0000	194	13.9284
45	6.7082	95	9.7468	145	12.0416	195	13.9642
46	6.7823	96	9.7980	146	12.0830	196	14.000
47	6.8537	97	9.8489	147	12.1244	197	14.0357
48	6.9282	98	9.8995	148	12.1655	198	14.0712
49	7.0000	99	9.9499	149	12.2066	199	14.1067
50	7.0711	100	10.0000	150	12.2474	200	14.1424

Study Cards

The following few pages contain useful information that you may want to remember. If you want to work on memorizing the multiplication table, the first few perfect squares, the first few primes, or other information, cut out the handy reminders from these pages and carry them with you in your pocket or purse. Refer to them at every opportunity. Quiz yourself on them. Get a friend or tutor to quiz you on them. Read and recite the material until you have it firmly in your memory.

ADDITION TABLE

+	0	1	2	3	4	5	6	7	8	9
0	0	1	2	3	4	5	6	7	8	9
1	1	2	3	4	5	6	7	8	9	10
2	2	3	4	5	6	7	8	9	10	11
3	3	4	5	6	7	8	9	10	11	12
4	4	5	6	7	8	9	10	11	12	13
5	5	6	7	8	9	10	11	12	13	14
6	6	7	8	9	10	11	12	13	14	15
7	7	8	9	10	11	12	13	14	15	16
8	8	9	10	11	12	13	14	15	16	17
9	9	10	11	12	13	14	15	16	17	18

MULTIPLICATION TABLE

×	0	1	2	3	4	5	6	7	8	9	10
0	0	0	0	0	0	0	0	0	0	0	0
1	0	1	2	3	4	5	6	7	8	9	10
2	0	2	4	6	8	10	12	14	16	18	20
3	0	3	6	9	12	15	18	21	24	27	30
4	0	4	8	12	16	20	24	28	32	36	40
5	0	5	10	15	20	25	30	35	40	45	50
6	0	6	12	18	24	30	36	42	48	54	60
7	0	7	14	21	28	35	42	49	56	63	70
8	0	8	16	24	32	40	48	56	64	72	80
9	0	9	18	27	36	45	54	63	72	81	90

PERFECT SQUARES

$1^2 = 1$	$6^2 = 36$	$11^2 = 121$	$16^2 = 256$
$2^2 = 4$	$7^2 = 49$	$12^2 = 144$	$17^2 = 289$
$3^2 = 9$	$8^2 = 64$	$13^2 = 169$	$18^2 = 324$
$4^2 = 16$	$9^2 = 81$	$14^2 = 196$	$19^2 = 361$
$5^2 = 25$	$10^2 = 100$	$15^2 = 225$	$20^2 = 400$

PRIMES LESS THAN 100

2	3	5	7	11
13	17	19	23	29
31	37	41	43	47
53	59	61	67	71
73	79	83	89	97

If	$A \times B = C$
then	$B = C \div A$
and	$A = C \div B$

Percent	Decimal	Fraction	Percent	Decimal	Fraction
5%	0.05	$\frac{1}{20}$	50%	0.50	$\frac{1}{2}$
$6\frac{1}{4}\%$	0.0625	$\frac{1}{16}$	60%	0.60	$\frac{3}{5}$
$8\frac{1}{3}\%$	$0.08\overline{3}$	$\frac{1}{12}$	$62\frac{1}{2}\%$	0.625	$\frac{5}{8}$
10%	0.10	$\frac{1}{10}$	$66\frac{2}{3}\%$	$0.6\overline{6}$	$\frac{2}{3}$
$12\frac{1}{2}\%$	0.125	$\frac{1}{8}$	70%	0.70	$\frac{7}{10}$
$16\frac{2}{3}\%$	$0.1\overline{6}$	$\frac{1}{6}$	75%	0.75	$\frac{3}{4}$
20%	0.20	$\frac{1}{5}$	80%	0.80	$\frac{4}{5}$
25%	0.25	$\frac{1}{4}$	$83\frac{1}{3}\%$	$0.8\overline{3}$	$\frac{5}{6}$
30%	0.30	$\frac{3}{10}$	$87\frac{1}{2}\%$	0.875	$\frac{7}{8}$
$33\frac{1}{3}\%$	$0.3\overline{3}$	$\frac{1}{3}$	90%	0.90	$\frac{9}{10}$
$37\frac{1}{2}\%$	0.375	$\frac{3}{8}$	100%	1.00	$\frac{10}{10}$
40%	0.40	$\frac{2}{5}$			

Signal Words	Translate as
Is, is equal to, equals, the same as	=
Of, the product of, multiply, times, multiplied by	×
Add, in addition, plus, more, more than, sum, and, increased by, added to	+
Subtract, subtract from, less, less than, difference, diminished by, decreased by	−
Divide, divided by	÷
Twice, double, twice as much	2×
Half of, half	½×

THE MOST OFTEN USED SQUARE ROOTS

$\sqrt{2} \approx 1.4142 \approx \dfrac{7}{5}$ or, even closer, $\dfrac{17}{12}$

$\sqrt{3} \approx 1.7321 \approx \dfrac{7}{4}$ or, even closer, $\dfrac{19}{11}$

$\sqrt{5} \approx 2.2361 \approx \dfrac{9}{4}$

$\sqrt{6} \approx 2.4495 \approx \dfrac{22}{9}$

$\sqrt{7} \approx 2.6457 \approx \dfrac{8}{3}$

$\sqrt{8} \approx 2.8284 \approx \dfrac{14}{5}$ or, even closer, $\dfrac{17}{6}$

$\sqrt{10} \approx 3.1623 \approx \dfrac{19}{6}$

Index

LaVergne, TN USA
03 May 2010
181364LV00005B/232/P